Soft Computing in Design and Manufacturing of Composite Material

Related titles:

Failure mechanisms in polymer matrix composites
(ISBN 978 1 84569 750 1)

Materials, design and manufacturing for lightweight vehicles
(ISBN 978 1 84569 463 0)

Introduction to aerospace materials
(ISBN 978 1 85573 946 8)

Soft Computing in Design and Manufacturing of Composite Material

Applications in brake friction and thermoset matrix composites

Dragan Aleksendrić and
Pierpaolo Carlone

ELSEVIER

AMSTERDAM • BOSTON • CAMBRIDGE • HEIDELBERG • LONDON
NEW YORK • OXFORD • PARIS • SAN DIEGO
SAN FRANCISCO • SINGAPORE • SYDNEY • TOKYO
Woodhead Publishing is an imprint of Elsevier

WP
WOODHEAD
PUBLISHING

Woodhead Publishing is an imprint of Elsevier
80 High Street, Sawston, Cambridge, CB22 3HJ, UK
225 Wyman Street, Waltham, MA 02451, USA
Langford Lane, Kidlington, OX5 1GB, UK

British Library Cataloguing-in-Publication Data
A catalogue record for this book is available from the British Library

Library of Congress Control Number: 2014948681

ISBN 978-1-78242-179-5 (print)
ISBN 978-1-78242-180-1 (online)

For information on all Woodhead Publishing publications visit our website at http://store.elsevier.com/

Typeset by RefineCatch Limited, Bungay, Suffolk

Printed in the United States of America

Contents

List of figures

List of tables

Preface

The remarkable potential and significant advantages of the application of advanced computational techniques has been highly recognized in the last decade. In particular, evident benefits can be achieved by applying these techniques in engineering fields characterized by high technological content, such as the design and optimization of the properties of composite materials. There is no doubt that the engineering of composite materials needs an integration of different techniques and methods. Many of these techniques are recognized as soft computing techniques. The main aim of soft computing is to explore different alternative designs that may satisfy imposed requirements or to find the optimal solution regarding a function goal and prescribed boundary conditions. It is also well known that the properties of the final product rely on the raw materials and on the manufacturing processes and conditions employed. In this framework, soft computing could be used effectively to find an individual trade-off between the formulation of a composite material and the manufacturing parameters against the final characteristics. It should also be kept in mind that this trade-off cannot exclude proper consideration of the specific operation conditions of the final product. In this sense, soft computing techniques provide intriguing tools in the area of composite material development and should be integrated into a system approach to the engineering of composite materials. Indeed, this system approach to the engineering of composite materials offers new possibilities not only for finding the best solutions out of many potentially valid alternatives, but also for creating new alternatives using the intelligent abilities of system/process models. Of course, a mandatory step for achieving the final goal is the development of computer-based models representing the properties and performance of a composite material, correlated with the formulation and manufacturing process. This book introduces soft computing as a method for intelligent engineering of the design and manufacture of composite materials, with the main aim being to highlight the potential product/process improvements achievable in advanced material optimization and development. The focus of the book is related to the investigation of the possibilities for development of intelligent models of the performance of composite materials, to be used for the optimization of the manufacturing process of such materials.

The book is divided into five chapters. The first chapter elaborates on the basic characteristics of composite materials and their classification. The second chapter explains why the engineering of composite materials should be integrated with the application of soft computing techniques. This chapter contains explanations of a system approach to the engineering of composite materials. The third chapter provides a description of the processes commonly employed for manufacturing advanced composite materials. In particular, the first part of the chapter provides a general overview of the manufacture of thermoset matrix composites, while the second part is more oriented towards materials suitable for the realization of a specific type of mechanical component, namely brake friction material composites. The basic characteristics of the manufacturing processes for the composite materials considered, and relevant issues, are also elaborated in this chapter. In the fourth chapter, the important abilities of computing techniques (artificial neural networks and genetic algorithms) are explained, as well as the possibilities for integration of soft computing techniques. Furthermore, other computational methods (FEM, FVM, FDM) that could be integrated with soft computing techniques are reported. The fifth chapter shows several different ways of using soft computing techniques, integrated with other computational methods, in the modelling, prediction and optimization of composite materials. The possibilities of soft computing techniques are demonstrated concerning the modelling, prediction and optimization of the performance of brake friction materials, as well as of thermoset matrix composites.

About the authors

Dragan Aleksendrić is an Associate Professor at the Automotive Department, Faculty of Mechanical Engineering, University of Belgrade. He received his Dipl. Ing. in Mechanical Engineering in 1996, Master of Science degree in Mechanical Engineering in 2000 and Doctor of Science degree in Mechanical Engineering in 2007 from the Faculty of Mechanical Engineering, University of Belgrade. His research is in the field of tribology of friction systems, specifically brakes and clutches, including R&D on brake systems and brake friction materials; system engineering; the application of artificial intelligence in the modelling, prediction and control of the performance of motor vehicle braking systems; vehicle safety; and type approval of braking systems for road vehicles and their components. He is Head of the Laboratory for Motor Vehicle Safety at the Faculty of Mechanical Engineering, University of Belgrade. He is an author or co-author of 15 chapters in national and international monographs, of more than 45 papers presented at international conferences, of 20 papers published in international journals, and of 10 technical solutions, and is a co-author of four software packages.

Affiliation: Automotive Department, Faculty of Mechanical Engineering, University of Belgrade, Serbia. Email: daleksendric@mas.bg.ac.rs.

Pierpaolo Carlone received his MS degree cum laude in Mechanical Engineering from the University of Salerno, Italy, in 2003 and his PhD in Mechanical Engineering in 2007 from the same institution. Currently he is Assistant Professor of Manufacturing Processes and Technology at the Department of Industrial Engineering of the University of Salerno. He is a member of the Scientific Committee of The European Association for Material Forming. His research activity is focused on the modelling, optimization and control of composite-material manufacturing processes and reliability analysis. He is author of about 70 papers published in refereed journals and conference proceedings.

Affiliation: Department of Industrial Engineering, University of Salerno, Italy. Email: pcarlone@unisa.it.

1

Introduction to composite materials

DOI: 10.1533/9781782421801.1

Abstract: Composite materials engineering needs systematic and interactive approaches, which should allow the achievement of optimum material characteristics. This process requires the application of various methods and technologies aiming at (i) investigation of the physical and mechanical properties of each constituent, as well as of the composite material; (ii) optimization of the properties of the composite according to the specific working conditions; (iii) understanding the effects of manufacturing and composition on the properties of the composite material; and (iv) development of computational methods for characterization, analysis and prediction of the performance of materials under different working conditions. This appears especially important in the case of composite materials characterized by strongly inhomogeneous properties.

Key words: composite materials, classification, properties.

1.1 Basics of composite materials

The 'composite' concept is not a human invention. Wood is a natural composite material consisting of one species of polymer – cellulose fibres with good strength and stiffness – in a resinous matrix of another polymer, the polysaccharide lignin [1]. The idea of combining different materials in order to take advantage of the peculiar positive properties of each single constituent has ancient historical roots. However, it is only in the last half century that the science and technology of composite materials have developed sufficiently to provide engineers with a novel class of materials and the necessary tools to enable them to use them advantageously [1]. A structural composite is a material system consisting of two or more phases on a macroscopic scale, whose mechanical performance and properties are designed to be superior to those of the constituent materials acting independently [2]. Generally, a composite material is composed of a reinforcement (fibres, particles and/or fillers) embedded in a matrix (a polymer, metal or ceramic). In general, the reinforcement is supposed to be the strongest of all of the components, and therefore, in structural composites, its main role is to enhance the overall mechanical properties of the resulting material.

The matrix is required to fulfil several functions, most of which are vital to the performance of the material. As a first approximation, it can be stated that the matrix holds the reinforcement in place to define the shape of the final product. The roles of the matrix in fibre-reinforced and particulate composites are quite different. The binder for a particulate aggregate simply serves to retain the composite mass in a solid form, but the matrix in a fibre composite performs a variety of other functions, which must be appreciated to understand the true composite action which determines the mechanical behaviour of a reinforced material. These functions should therefore be considered in some detail [1]. The matrix binds the fibres together, holding them aligned in the important stressed directions. Loads applied to the composite are then transferred into the fibres, the principal load-bearing component, through the matrix, enabling the composite to withstand compression, flexural and shear forces as well as tensile loads. The ability of composites reinforced with short fibres to support loads of any kind is dependent on the presence of the matrix as the load transfer medium, and the efficiency of this load transfer is directly related to the quality of the fibre/matrix bond. The matrix must also isolate the fibres from each other so that they can act as separate entities. Many reinforcing fibres are brittle solids with highly variable strengths. When such materials are used in the form of fine fibres, not only are the fibres stronger than the monolithic form of the same solid, but there is also the additional benefit that the fibre aggregate does not fail catastrophically. The matrix should protect the reinforcing filaments from mechanical damage (abrasion) and from environmental attack. Through the quality of its 'grip' on the fibres (the interfacial bond strength), the matrix can also be an important means of increasing the toughness of the composite [1].

Composites have unique advantages over monolithic materials, such as high strength, high stiffness, long fatigue life, low density and adaptability to the intended function of the structure. Additional improvements can be realized in terms of corrosion [2]. The simple term 'composites' gives little indication of the vast range of individual combinations that are included in this class of materials. Polymers are often two-phase composites, consisting of a matrix of one polymer with distributions of harder or softer particles contained within it; wood is a perfect example of this. Concrete (the direct descendant of straw and mud bricks) is a classic example of a ceramic/ceramic composite, with particles of sand and aggregate of graded sizes in a matrix of hydrated Portland cement. These materials have been well known for many years, and materials scientists have learned to control their properties by controlling their microstructures; that is to say, the quantity, the shape and the distribution of what we might refer to as the 'reinforcing phase'. The idea of mixing components across materials class boundaries is a natural extension of this idea [1].

Composites are commonly classified at two distinct levels. The first level of classification is usually with respect to the matrix constituent. The major composite classes include organic-matrix composites (OMCs), metal-matrix composites (MMCs) and ceramic-matrix composites (CMCs). The term 'organic-matrix composite' is generally assumed to include two classes of composites: polymer-matrix composites (PMCs) and carbon-matrix composites (commonly referred to as carbon–carbon composites) [3]. The second level of classification refers to the form of the reinforcement: particulate reinforcements, whisker reinforcements, continuous-fibre laminated composites and woven composites [3]. Composites are used not only for their structural properties but also for electrical, thermal, tribological and environmental applications. These features rely strongly on the specific constituents combined in the composite, on the extent of their presence in the final material (weight or volume fraction), and on the shape and architecture of the reinforcing phase. Ideally, the properties of engineering materials should be reproducible and accurately known. Since satisfactory exploitation of the composite principle depends on the design flexibility that results from tailoring the properties of a combination of materials to suit a particular requirement, the accurate prediction of those properties is imperative. At the present time, some of the more important engineering properties of composites can be estimated well on the basis of mathematical models, but many cannot [1].

What is more, modern composite materials are usually optimized (with respect to the aforementioned aspects) to achieve a particular balance of properties for a given range of applications [3]. However, as a common practical definition, the term 'composite materials' may be restricted to emphasize those materials that contain a continuous matrix constituent that binds together and provides form to an array of a stronger, stiffer reinforcement constituent. When designed properly, the new combined material exhibits better strength than would each individual material. The main advantages of composite materials are their high strength and stiffness, combined with low density, when compared with bulk materials, allowing weight reduction in the finished part [4]. In composites, materials are combined in such a way as to enable us to make better use of their virtues while minimizing to some extent the effects of their deficiencies. This process of optimization can release a designer from the constraints associated with the selection and manufacture of conventional materials. The designer can make use of tougher and lighter materials, with properties that can be tailored to suit particular design requirements [1].

The obtainment of tailored and desired properties cannot preclude specific manufacturing considerations. Indeed, the need to combine different materials appropriately and realize a well-defined product has a remarkable impact on the manufacturing processes that are suitable. Limiting these considerations to the most widely used composite materials, i.e. PMCs, the aforementioned combination can be realized offline or online: in the former case, two different processing steps, given by the combination of the constituents and the shaping and solidification of the product, can be identified, whereas, in the latter case, the two operations are performed more or less simultaneously. As generally happens for all engineering materials, the early choice of the manufacturing process is based on the dominant shape of the product and cost considerations; however, in the case of fibrous composites, the architecture of the reinforcement can also be a key issue to consider. What is more, great attention should be given to the planning of the process, i.e. the definition of suitable process parameters. Indeed, even assuming that constituent materials and their volume fractions are correctly defined, the use of erroneous (or at least non-optimized) process set-ups could drastically increase the content of voids or negatively affect the fibre–matrix interface, inducing a global worsening of performance.

1.2 Conclusions

Further technological development depends strongly on advances in the field of materials. Composite materials represent a new approach in the process of materials development. Combining two or more materials to obtain a new material with enhanced properties offers wide possibilities for improvements. Optimization of a composite material, regarding not only its matrix and reinforcement materials but also the manufacturing process, could provide a new material that meets previously defined requirements for a specific application. Additionally, composites provide intriguing advantages in terms of design flexibility, because many of them can be manufactured into complex shapes. A strong impulse in this direction could be given by the application of appropriate techniques for modelling, prediction and optimization of the properties of composite materials, as well as of the inherent manufacturing processes.

1.3 References

[1] Harris B. *Engineering Composite Materials*, The Institute of Materials, London (1999).

[2] Isaac M.D., Ori I. *Engineering Mechanics of Composite Material*, Oxford University Press (1994).

[3] Miracle D.B., Donaldson S. (eds) *ASM Handbook*, Volume 21: *Composites*, ASM International (2001).

[4] Campbell F.C. *Introduction to Composite Materials*, ASM International (2010).

2

Soft computing and composite materials engineering

DOI: 10.1533/9781782421801.7

Abstract: Engineers and researchers have to develop and use sophisticated tools, methods and technologies in order to enable advanced technical systems to provide customer satisfaction and to become successful in the marketplace. Soft computing can be considered as a new tool of systems engineering. The systems approach to problem solving in engineering needs advanced 'top-down' methods, particularly at the design stage. Since a systems approach is needed in the case of composite materials engineering, soft computing techniques could be used in different phases of the engineering of a composite material. It is especially related to functional analysis, design, testing, prediction and optimization. The integrated process of modelling, calculation, prediction and/or optimization of the properties of a composite material against different influencing factors is the most important part of materials engineering. That is why soft computing techniques, integrated with different numerical methods, could support that process.

Key words: soft computing, composite materials, systems engineering.

2.1 Introduction

Industry today, especially the automotive industry, demands products characterized by excellent quality with shorter time to market. That is why engineers have to use advanced technologies. In order to enable advanced technical systems to provide customer satisfaction and to become successful in the marketplace, it is necessary to apply updated technologies or technical skills and methods in all phases of concept analysis, design, development, simulation, testing, manufacturing, verification, certification, marketing and service. This is a result of the designer's intention, which asserts that quality should be 'inbuilt' in a system, not merely 'added' to it. New technologies should enable engineers to fulfil all of the requirements imposed on technical systems by their users and by governmental or supra-governmental authorities. This means that tuning of mutually opposing, and often contradictory, requirements is necessary. That is why advanced techniques of systems approaches are needed. Composite materials engineering also needs a systems approach which can integrate different tools, including soft computing techniques.

It is known that systems engineering is not a single engineering discipline. It comprises the efforts necessary to integrate different product characteristics into the total engineering effort, and it applies updated computer- and model-based engineering tools to assist in the optimization of the product design, reducing development time. There is an emerging awareness among professionals about systems engineering and systems thinking. It also integrates related technical parameters and assures compatibility of all physical, functional and program interfaces in a manner that optimizes the total definition and design of the system. It transforms an operational need into a description of the performance parameters of a system and a preferred system configuration through the use of an iterative process of functional analysis, synthesis, optimization, definition, design, testing and evaluation. By definition, systems engineering delivers both a product and a process for manufacturing it, and it relates to the full system life cycle or the consumer-to-consumer cycle, namely to the development, manufacture and usage of a system. It is a 'top-down' method in which a variety of engineering techniques and technologies related to the system life cycle are applied simultaneously and associated with each other, with the aim of fulfilling a common task. Consequently, the process of composite materials engineering needs a proper combination of methods and techniques for the successful development of the materials. The approach to composite materials engineering should be based on integrated engineering – a new concept integrating ideas, methodologies and tools. Obviously, soft computing techniques need to be incorporated into the process of composite materials engineering.

2.2 Soft computing and systems engineering

Systems engineering efforts provide a conceptual framework for managing the life cycle of a system [1]. The effects of interactions among the many parts of the overall system must be taken into account during the engineering process. The system clearly defines the most desired result before any attempt is made to accomplish it [1]. The analysis for the detailed design phase in the systems engineering approach includes the testing and evaluation of prototype components and subsystems, because physical and functional interactions between subsystems of a given system are accounted for explicitly. Some system descriptors are purpose, function, process, boundaries, environments, dynamics, stability, adaptation, information, control, interactions, interfaces, inputs and outputs, and life cycle [1].

This approach starts with the problem formulation, the definition of system requirements and other specifications. It is good engineering practice to state the problem in terms of the top-level function that the system must perform, and this is often called the 'mission of the system'. Generally, it is said that the word 'optimal' should be used carefully in the problem formulation, because often there is no single optimal solution for a complex system problem. Systems engineering normally uses a set of alternative designs that satisfy the system requirements and specifications to varying degrees. There will always be some trade-offs, because moving from one alternative to another will probably improve at least one criterion and worsen at least one other. This is a point where soft computing could offer new possibilities not only for finding the best solutions out of many alternatives that could be considered but also for creating new alternatives using the intelligent capabilities of system/process models.

Systems engineering is similar to innovation in engineering. The initial step should be product identification or producing a product description. The subsequent steps may be (a) to design the product in detail, including analysis, synthesis, simulation and testing; (b) to identify and resolve any effects of interactions caused by innovations; (c) to develop a manufacturing process, incorporating innovations; (d) to develop a process for assuring quality in the new system; and (e) to develop plans and processes for marketing and servicing the system [1]. It is obvious that systems engineering includes problem formulation, mathematical or computer modelling of complex systems (with their physical, economic and social-preference variables), optimal decision making (perhaps with multiple criteria), decision making and prediction with uncertain information, and procedures for conceptualizing and controlling the behaviour of a system. In other words, soft computing could be used in several different phases of a systems approach, because many of the steps mentioned above need computer-based models, which could replace a physical prototype. Accordingly, the product development process shown in Figure 2.1 should be based on appropriate computer-based models for performing prediction and optimization of product performance.

The development process based on the systems approach should enable one (a) to verify that the system as a whole satisfies all its system-level requirements; (b) to trace all shortcomings to their source; (c) to measure all significant characteristics of the as-built system, especially those for which new models need to be developed; and (d) to validate and/or calibrate the models used in the development process so that the effect of future modifications can be predicted with confidence. If further optimization is needed, the testing phase is repeated; if not, the final component and subsystem designs are chosen and integrated into the final system design. Engineers seek appropriate methods and techniques for these development activities. Many of these techniques belong to the so-called soft computing techniques and represent an important part of a systems approach. Systems engineering is known as a 'top-down' approach to the life cycle of a product, enabling the best way of 'confrontation' of all mutually opposed requirements, so as to find an 'optimized' solution to the problem by their 'simultaneous' satisfaction. While the systems engineering approach replaces 'cut-and-try' engineering by modelling and analysis techniques and reduces the development time and cost of the system, providing a quality product that meets or exceeds

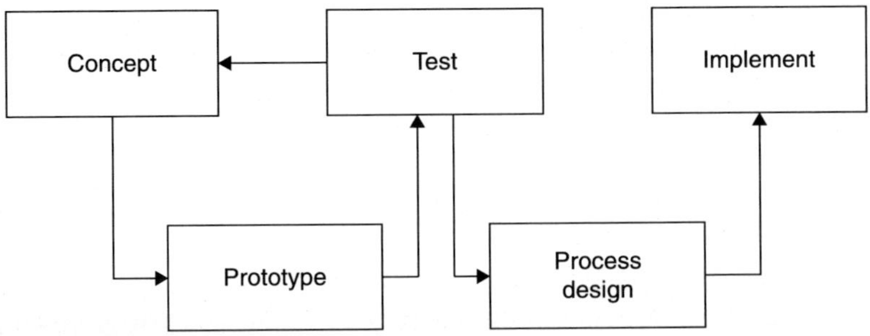

Figure 2.1 The process of product development

customers' requirements, too many iterations of the process of testing and validation of physical prototypes will not favour reaching that goal adequately.

Like other complex mechanical systems, the braking system of a road vehicle, for instance, depends on a hierarchy as shown in Figure 2.2. This complex system is defined by subsystems (two independent service braking systems, an emergency braking system and a parking braking system), assemblies (control systems, transmission and wheel brakes, for example), subassemblies (calipers including a brake piston and wear compensation device, in addition to friction pads, for example) and components (for example, a brake rotor, which may be a drum or a disc depending on the type of brake; a friction material, which may be a brake pad or a brake lining; and a brake control device, which may be a wheel cylinder). Figure 2.2 also shows that analysing only system-level requirements will not be sufficient to enable the identification of the mission imposed on the subsystem, assembly or individual components. Namely, each of them 'deserves' its own-level requirements, which must also depend on a hierarchy.

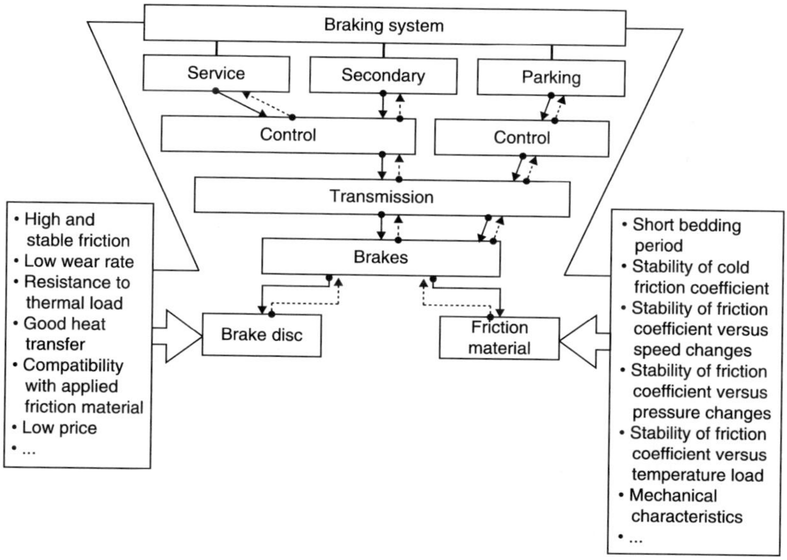

Figure 2.2 Top-down–bottom-up approach to the development of a braking system regarding the properties of a friction pair

If we take into consideration only the characteristics of the brake friction pair, the overall performance of the braking system can be affected in different ways by complex interactions between the brake disc and the brake friction material, which is a complex composite material. That is why the performance of a braking system should be modelled, predicted and optimized against these complex influences. This needs modelling, prediction and optimization of the performance of the brake components, such as the friction material. The modelling, prediction and optimization of the performance of a brake friction material are quite difficult tasks, which require implementation and integration of advanced methods such as finite element analysis, artificial neural networks and genetic algorithms.

The remarkable potential and significant advantages of the application of advanced computational techniques are also highly recognized during the design and optimization of the process parameters to be employed during the manufacturing stage. For instance, the curing process is a key step in the manufacturing of brake friction materials and thermoset matrix composites, independently of the technology used to shape the workpiece, and several related issues deserve proper consideration. In the case of a thermoset matrix composite, the temperature cycle imposed (the autoclave temperature or the die temperature used in pultrusion), the overpressure applied (the autoclave pressure or the liquid pressure used in pultrusion) and the possible vacuum facility define, in a complex way, the temperature history of different zones of the material during processing, the distribution of the degree of cure in the final product, the volume fractions, the percentage and size of voids, the residual stresses, and the workpiece distortion [2]. As a consequence, the influence of the process parameters is not limited to the performance of the process in terms of productivity, energy intensity or production cost, but concerns also the quality and mechanical properties of the final product. The cycle design, i.e. the definition and optimization of the temperature–time curve, is a crucial issue for a competitive production process. Today, industrial practice is still based to a remarkable extent on trial and error procedures to accomplish this task. These methods, however, are generally expensive in time and money and, considering the complexity of the problem, do not ensure the individuation of optimal or suboptimal solutions. During the curing process, the thermal energy provided to the composite increases the temperature of the part, resulting in an initial decrease in the viscosity of the material and in activation of the curing reaction. Then, cross-linking of the resin monomers induces an increase in the degree of cure and viscosity until solidification of the part. The cross-linking process, in conjunction with the existing temperature field, is characterized by two relevant phenomena, namely gelation and vitrification, in which the state of the resin changes, affecting resin flow, the compaction of the material and the mechanical properties. Curing processes characterized by a faster heating ramp and a higher holding temperature allow one to obtain a higher degree of cure in a shorter time; however, the risk of hot spots and resin degradation is quite relevant. What is more, undesired outside–in solidification, excessive residual stresses and distortions could be induced. On the other hand, slow heating increases the cycle time, and a low holding temperature could decrease the final degree of cure. Pressure and vacuum cycles need to be combined appropriately with heat input in order to maximize the achievable benefits in terms of removal of entrapped air, gases and excess resin, as well as to improve the dimensional precision of the workpiece. The complex phenomena that accompany the manufacturing process of the composite material need to be carefully analysed, modelled, predicted and optimized. In this context, soft computing techniques could be employed effectively.

2.3 Conclusions

The benefits provided by the development and implementation of suitable modelling, prediction, simulation and optimization tools are quite evident. In particular, the opportunities for the adoption of simulation-based optimization procedures seem promising, owing to the numerous potential solutions that can be considered until convergence is achieved. Integrated systems, based on a combination of soft computing techniques, could be applied effectively to the modelling, prediction and optimization of processes, as well as for their monitoring and control. In this framework, artificial-neural-network-based models reinforced with genetic algorithms and other computational methods could create an integrated engineering environment for the design and development of composite materials.

2.4 References

[1] Aleksendrić D., Duboka Č. Virtual reality – new technology for systems engineering, *FISITA World Automotive Congress*, Paris, F98T667 (1998).

[2] Gutowski T.G. *Advanced Composites Manufacturing*, Wiley Interscience, New York (1997).

Composite materials manufacturing

DOI: 10.1533/9781782421801.15

Abstract: The manufacturing process is very important with regard to the final properties of a composite material. The manufacturing processes employed to realize brake friction materials or thermoset matrix composites have a crucial impact on their future properties. This is especially related to the level and stability of friction and wear during braking in the case of brake friction materials. The development of a composite material is strongly affected by its formulation and its manufacturing conditions. Owing to the complex and interrelated influences of the formulation and the manufacturing conditions, it is difficult to find the best set of process parameters for a specific material formulation. Accordingly, the selection, mixing and preparation of raw materials, as well as the choice of moulding pressure, moulding time, moulding temperature, heat treatment time and/or heat treatment temperature, can be done over a wide range. In this chapter, the basic characteristics of the manufacturing processes for brake friction materials and thermoset matrix composites are elaborated on.

Key words: manufacturing process, thermoset matrix composite, brake friction material, manufacturing conditions, disc pad.

3.1 Manufacturing of thermoset matrix composites

3.1.1 *Introduction*

In recent years, polymeric matrix composite materials have been widely used for several applications in different fields, such as automotive, aerospace, aeronautical, nautical, energy and sporting goods. The demand for high performance and constant quality, combined with the need to reduce costs arising from any possible manufacturing inefficiencies, has resulted in the increasing use and development of industrial manufacturing processes characterized by low human intervention [1]. In several cases, a reduction in human presence is also strongly desirable to avoid health hazards caused by the emission of volatiles during the resin reaction, such as, for instance, styrene emission during the curing process of polyester resins [2].

As a general principle, the manufacturing process employed to realize (shape, cure or consolidate) a polymeric matrix composite material should be able to:

- respect the geometrical shape, dimensions, tolerances and surface finish of the product;
- realize the desired reinforcement architecture (long, continuous fibres), orientation (short, discontinuous fibres) or distribution (particulate), ensuring the correct reinforcement volume fraction as defined at the design stage;
- promote complete wetting of the reinforcement and the development of a fibre–matrix interphase, able to transfer loads but also to allow damage mechanisms, such as fibre debonding or pull-out, to dissipate energy;
- achieve a satisfactory distribution of the degree of cure (for thermoset resins) or degree of crystallization (for thermoplastic resins), within a reasonable processing time, avoiding resin degradation or excessive residual stresses;
- keep the porosity (voids) content at an acceptable level, generally defined considering the applicable sector;
- minimize the production cost.

The choice of the manufacturing process, as well as, obviously, of the processing dynamics and parameters, is significantly affected by the resin system to be used as the matrix, distinguishing first of all the processing of thermosets (epoxies, polyesters, vinyl esters, bismaleimides etc.) from that of thermoplastics (PEEK etc.). Thermoplastic resins are generally characterized by a higher molecular weight at the beginning of the manufacturing process (impregnation stage), showing viscosity values several order of magnitude higher than those of epoxies and polyester resins. As a consequence, the flow of thermoplastics in fibre beds is quite difficult and requires higher processing temperatures and pressures to ensure satisfactory fibre wetting. Then, solidification of the material is induced by cooling, being a crystallization phenomenon promoted by temperature reduction, where the local cooling rate significantly affects the final degree of crystallization. For this reason, in several cases where thermoplastic composite materials characterized by continuous reinforcement are processed, the matrix–fibre combination step is separated from the subsequent tasks. Different approaches are possible when one is manufacturing short fibres or particulate reinforcements.

Generally speaking, matrix–fibre combination can be performed offline or online. In the former case, the combination of the reinforcement and matrix, and possibly the creation of the desired interphase, anticipates the shaping and consolidation of the product (as above). These two tasks are generally well separated and can be performed in different industrial plants. Typical examples are pre-impregnated layers (pre-pregs) and commingled yarns (where the matrix and reinforcement are in the form of braided fibres, reducing the distance that has to be covered by the resin flow to create a consistent interphase). In the latter case, the fibre wetting and product shaping are done at the same time, as for instance in liquid composite moulding processes, or as consecutive tasks on the same machine (filament winding) or the same production line (pultrusion).

This procedure is commonly adopted when thermoset composite materials are processed, thanks to the low values of molecular weight and viscosity shown in the early stages of the manufacturing process. Differently from thermoplastics, the shaping of thermosets is performed at relatively low temperatures, in order to prevent early activation of the cure reaction and gelation of the resin system, which is an undesired phenomenon during the impregnation of the fibre bed, since it is accompanied by a sharp viscosity increase, inhibiting matrix flow. Subsequently, the impregnated but not consolidated composite undergoes a heating (curing) cycle: the heat transferred to the resin activates, accelerates and promotes the cross-linking of the monomers, increasing the degree of polymerization, the molecular weight and the viscosity until solidification of the part. A few thermoset resin systems, however, such as the Avimid N polyimide resin, do not follow this basic principle. Indeed, the high viscosity values exhibited even at a very low polymerization degree impose the adoption of processing parameters, in terms of temperature and pressure, closer to the ones used for the manufacture of thermoplastic rather than thermoset composites.

The rheological behaviour of the resin to be processed, i.e. the relation between viscosity, temperature and degree of cure, plays a key role in the manufacturing of thermoset composites. In particular, viscosity tends to decrease with increasing temperature and to increase with the degree of cure. It should be noted that the overall balance is absolutely not trivial, since the thermal energy provided to the composite material increases the material temperature, resulting initially in a decrease in the material viscosity. However, it also activates the cure reaction, inducing an opposite effect. As a consequence, at the very beginning of the manufacturing process, the heat input lowers the viscosity, resulting in a beneficial effect with respect to the impregnation of the reinforcement, the shaping of the workpiece, and the removal of entrapped volatiles and excess resin. Subsequently, the interconnection of the polymeric chains due to the exothermic cure reaction prevails over the molecular mobility, until gelation of the part and then solidification. In Figures 3.1 and 3.2, the profiles of the degree of cure and viscosity experienced by a Shell Epon 9420/9470/537 resin subjected to the temperature cycle indicated are depicted. These contrasting effects define the process window for a thermoset matrix composite, i.e. the time available for proper impregnation, shaping and void removal before material consolidation.

The above profiles can differ significantly between locations in the material to be processed for several reasons, such as local variations in the workpiece thickness or the volume fractions, as well as the influence of the tooling and the applied loads. This aspect is of great importance when planning a generic composite-manufacturing process since it directly affects the final product. Resin-rich zones are generally more

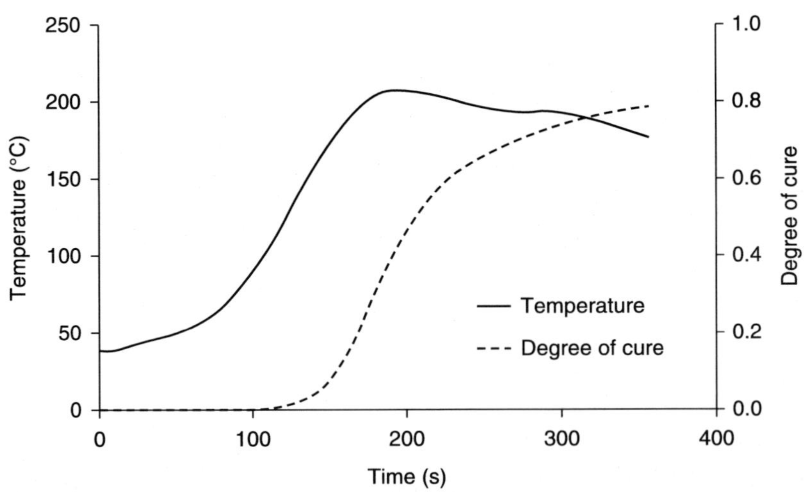

Figure 3.1 **Temperature and degree-of-cure profiles for a Shell Epon 9420/9470/537 resin**

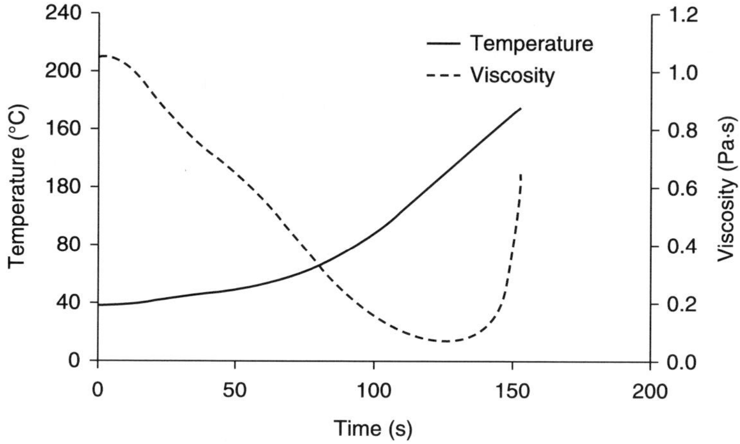

Figure 3.2 **Temperature and viscosity profiles for a Shell Epon 9420/9470/537 resin**

sensitive to thermal runaway and hot spots, owing to the heat released by the cure reaction. Furthermore, gradients in the temperature and degree of cure directly influence the transient and final stress distributions, inducing undesired residual scenarios, in particular in the case of outside–in solidification. This is generally caused by excessively fast activation of the cross-linking reaction on the workpiece surface with respect to the core, and is usually attributable to non-optimal design of the curing cycle.

Resin sellers provide suitable curing cycles for the resin/catalyser/accelerator systems that they supply; but, obviously, this information is not case-specific. In other words, they can indicate how (using what temperature and pressure cycles) the matrix should be cured, but not how to induce, to a reasonable approximation, the appropriate cycle in each location of the composite to be processed. In this context, the intriguing capabilities offered by computational techniques, such as simulation, inverse analysis and optimization tools, are simply amazing.

Since the main focus of this book is the application of soft computing techniques to thermoset composite design and manufacture, a complete and detailed description of all currently available processes for the manufacturing of composite materials has been omitted in the interests of brevity. For further and exhaustive reading, the references [3–7] cited at the end of this chapter are suggested. In the following, the manufacturing processes most commonly employed industrially for thermoset matrix composites are recalled. Techniques characterized by a large amount of human intervention, such as hand or spray lay-up are intentionally omitted, even if still widely adopted in the field. Manufacturing processes for short-fibre/particulate composite materials are described in the next section, focusing attention on the particular case of brake friction materials.

3.1.2 Autoclave processing

Generally speaking, the manufacturing process of a thermoset resin involves a preforming step followed by thermal curing of the workpiece and inspection/reworking/assembly operations. In several cases, preformed parts are vacuum bagged and subjected to a specific curing process, based on pressure and temperature cycles and using a vacuum facility, to obtain mechanical properties enhanced by up to 20–30% compared with other curing methods that do not involve external pressure on the workpiece. The increase in mechanical properties can be related to three aspects: material consolidation (thickness reductions are accompanied by an increase in the reinforcement volume fraction), full polymerization of the resin, and void reduction or suppression. This manufacturing step is commonly adopted also when high-temperature resin systems are processed. An autoclave is a pressure vessel equipped with a heating system. The term 'autoclave' is derived from the combination of two Greek words, *auto* (self) and *clavis* (closure), and refers to the capability of the first autoclaves to hermetically lock themselves, thanks to the inside–out (overpressure versus atmospheric) pressure. Autoclaves are usually configured as closed-end cylindrical tubes with an opening door at one end. The dimensions, and obviously costs, are extremely variable, depending on the specific production process. The external diameter and the length of large autoclaves (used to cure or post-cure workpieces with very large dimensions or to manufacture several small parts simultaneously) can reach about 25 and 50 m, respectively.

A generic autoclave curing cycle consists of one or more heating/holding steps followed by a final (slow) cooling to room temperature. After material positioning, the autoclave temperature is gradually increased by means of hot recirculating air or other gases (nitrogenous, for instance), at a rate defined taking into account the thickness and the physical properties of the part to be processed. At the same time, the vacuum and pressure facilities of the autoclave are activated. For thicker products, one or more temperature-holding steps are commonly introduced. The thermal energy provided to the composite increases the temperature of the part, resulting initially in a decrease in material viscosity, but also activation of the cure reaction. The viscosity reduction allows excess resin and entrapped volatiles to escape from the impregnated reinforcement, under the action of the autoclave pressure and the applied vacuum. On the other hand, a direct consequence of the cure reaction is an increase in the degree of cure and, consequently, in the viscosity, preventing further compaction, until the solidification of the part. It should be also noted that the applied pressure is generally carried by both constituents of the composite, i.e. the matrix and the reinforcement. This aspect plays an important role in relation to the evolution of voids. Indeed, in the case of excessive compaction due to resin removal, the benefits gained from the increase in fibre volume fraction can vanish because the reduction in the resin pressure acts as a driving force for void growth. The internal evolution of the cure reaction deserves specific attention, too. The solidification direction is strictly related to the activation of the resin reaction. In conventional curing systems such as autoclave curing, this aspect typically relies on conductive heat fluxes. As a consequence, an earlier cross-linking is initiated on the surface of the part; however, in the core of the material, the stronger heat generation increases the local temperature and results in a faster reaction. The intersection point of the cure profiles (at the surface and in the core of the material) defines the 'cure crossover'. Taking into account that the mechanical properties improve substantially after the gel point (AGP), it follows that the cure crossover should be induced close to or at least before the gel point. An appropriate design of the cure process is crucial to obtaining the desired material shaping and consolidation, to removing resin excess and entrapped volatiles, and to reaching full cure of the material without any thermal degradation of the resin system or distortion of the part related to the excessive thermal stress [8, 9]. Moreover, the progression of the cure front strongly affects the residual stresses and distortions in the processed part, since inside–out solidification is preferable to outside–in solidification [10].

3.1.3 Liquid composite moulding processes

This class of composite-manufacturing processes is characterized by the impregnation of a dry fibrous preform, appropriately placed between solid and solid or solid and flexible dies, by means of injection or infusion of the catalysed resin. In recent years, liquid composite moulding (LCM) processes have gained a lot of attention in industry, owing to the need to reduce human exposure to dangerous volatiles released during the impregnation and cure processes. The main advantages of LCM processes are related to the capability to manufacture products characterized by complex double-curvature shapes with remarkable precision and flexibility of the reinforcement architecture, to the relatively high and uniform reinforcement volume fraction, to the good surface finish, and to the reduced contact of operators with the reactive resin.

Several variants of the LCM process, such as the resin transfer moulding (RTM), vacuum-assisted resin transfer moulding (VARTM), resin infusion under flexible tool (RIFT), resin infusion under double flexible tool (RIDFT), resin film infusion (RFI), Seemann's composite resin infusion moulding process (SCRIMP) and flow flooding chamber (FFC) processes, are currently employed or under development. The differences between the aforementioned processes are related to the die material types (solid–solid, solid–flexible or flexible–flexible), to the initial state of the resin (catalysed liquid or film), and to the driving force inducing the resin to flow and to impregnate the preform (external overpressure or internal vacuum).

In particular, the RTM process is relatively efficient and automated, and is widely used for medium to high production volumes [11]. The process is based on several consecutive steps, consisting of preparation of a fibre preform characterized by the desired architecture, positioning of the preform into an appropriately shaped solid die, closure of the die by means of a rigid counter-die, injection of the resin, heating of the assembly, resin reaction and solidification of the workpiece, cooling, and extraction of the final product. A critical aspect of the process is related to the die design, taking into account injection gates, air vents, and heating and cooling systems. The VARTM process is similar to the RTM process; however, in this case, a vacuum pressure instead of a positive pressure is applied to promote preform impregnation [12].

In the RIFT process [13], the rigid counter-die is replaced by a flexible die, obtained using a vacuum bag, resulting in lower cost of the tooling with respect to the conventional RTM process, while in the RIDFT process [14], the use of flexible–flexible dies allows impregnation of a planar preform, the desired product shaping, and finally the curing and solidification of the material. In resin infusion processes, a vacuum-driven resin impregnates an appropriately placed preform. Vacuum pressure is applied inside the bagged material by means of a vacuum pump, while a resin trap is interposed between the bag and the pump to collect excess resin. Spiral tubing and resin flow lines are widely used to accelerate the filling stage. Further improvements in the filling stage can be achieved by employing the SCRIMP or the FFC process. In the SCRIMP process [15], an appropriate distribution medium, characterized by relatively high permeability, is placed between the peel ply and the vacuum bag, allowing the resin to spread rapidly over the product surface and then to impregnate the preform by flowing through the thickness. In the FFC process [16], a similar resin flow is achieved by means of a rigid chamber placed on the vacuum bag. In the first stage of the process, a vacuum pressure is applied between the die and the bag, and an even higher vacuum is induced between the chamber and the bag; as a consequence, an empty space is created between the preform and the bag, allowing the resin to flow faster. Once the preform is covered by the resin, the vacuum inside the rigid chamber is released and atmospheric pressure pushes the resin into the preform. The RFI process is based on the use of a catalysed resin film interposed between the layers of fibres; all the material is vacuum bagged. The combined action of heat and pressure liquefies the resin, which flows through the preform, and activates the cure reaction [17].

3.1.4 Pultrusion

Pultrusion is a continuous manufacturing process used mainly to realize composite rectilinear profiles characterized by a constant cross-section, although recent advances in the development of the process have allowed manufacturers to overcome these limitations. During the last two decades, the pultrusion process has experienced remarkable growth within the composites industry, owing to its cost-effectiveness, the use of automation and the high quality of the resulting products. Nowadays, the process is widely used to manufacture highly strengthened structures such as wind turbine blades, window profiles, door panels and reinforcing bars for concrete. The aforementioned advantages, combined with the relatively low energy intensity compared with other composite-manufacturing processes (see Figure 3.3), strongly reduce the environmental impact of pultruded composite structures over their entire life cycle. For some applications in the automotive industry, this impact is estimated to be lower than that of other engineering materials [18].

A schematic view of the pultrusion process is depicted in Figure 3.4. During the process, the reinforcement fibres, in the form of rovings or mat, are pulled through guides and impregnated by the resin material in an open bath or by employing a resin injection chamber. Wetted-out reinforcements are then pulled via a pulling mechanism through a heating

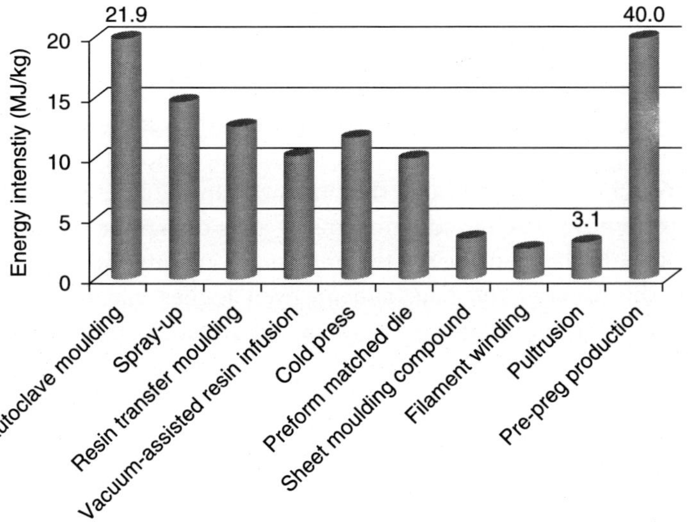

Figure 3.3 **Energy intensity of composite-manufacturing processes**

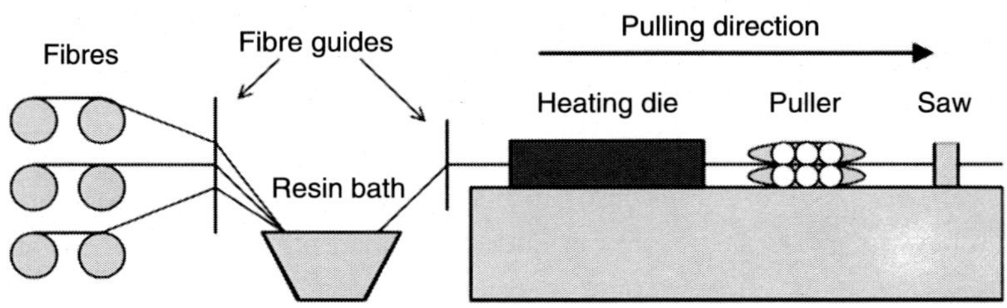

Figure 3.4 **Schematic view of the pultrusion process**

die. The die inlet is typically characterized by a tapered or conical convergent shape, in order to promote the desired impregnation and compaction of the reinforcement and the removal of air and excess resin. Furthermore, cooling channels are often machined into this portion of the die to prevent premature activation of the cure reaction and subsequent early gelation of the resin. In the straight portion of the die, heat provided by means of electrical heaters (cartridges, strips or platens) or hot oil activates the exothermic cure reaction of the thermoset resin [19]. As a consequence, the material changes its status from a reactive liquid to a gel and then a vitrified solid. At the end of the process, the cured and solidified product is pulled using reciprocating or caterpillar mechanical systems and then cut to the desired length.

The process is governed by several parameters, such as the resin bath temperature (or possibly the preheating temperature), the die temperature, the pull speed and the parameters of the cooling facility, which should be carefully defined, considering the geometry and dimensions of the part in conjunction with the thermochemical and rheological behaviour of the thermoset resin to be processed.

Even if the process is conceptually quite simple, the analysis of its dynamics and the definition of the optimal processing parameters is a complex task, owing to the mutual interactions between the physical phenomena involved, related mainly to the impregnation of the reinforcement (resin flow), heat transfer, species conversion and phase changes.

What is more, during the curing process, the resin is subjected to two different and contrasting phenomena, namely thermal expansion and chemical shrinkage, which influence the dimensions of the part, the status of the die–material contact, and the development of stresses, strains and distortions [20]. Indeed, in the first part of the die, the heat transferred from the die to the workpiece and the energy released by the exothermic reaction increase the temperature of the material, leading to its natural expansion. However, as the reaction proceeds, the cross-linking of monomers forces the resin to shrink, promoting contraction of the workpiece. When this latter phenomenon prevails, the part detaches from the die (in the case of full sections) and the contact pressure between the die and the composite vanishes.

Some variations of the conventional pultrusion process described above are needed when thermoplastic matrix composites are processed. In particular, since thermoplastics solidify on cooling, the heating portion of the die is connected to a cooling die to ensure gradual resin crystallization. Furthermore, commingled fibrous matrix or powders, instead of catalysed liquid, are used to overcome impregnation problems caused by the higher viscosity of thermoplastic resins [21, 22].

3.1.5 Filament winding

Roughly speaking, this process can be imagined as the reverse of the traditional turning process, where the difference lies in the fact that material is added to instead of removed from the part. During the filament-winding process, continuous reinforcement rovings or layers, impregnated with catalysed resin, are wound around a rotating mandrel, in order to obtain, at the end of the curing process (in an autoclave, for instance), hollow axisymmetric products. The pre-tensioning of the fibres imposes limitations on the feasible winding patterns; in other words, to avoid displacement of the fibres, a geodesic path must be followed. Two relevant features should be exhibited by a resin system for it to be effectively employed in a filament-winding process: an appropriate viscosity (1–2 Pa s) before the activation of the cure reaction, in order to wet the reinforcement, and a sufficiently long pot-life, in order to allow all the reinforcing layers to be deposited without any premature resin gelation. This technique offers intriguing advantages when circumferential, polar or helical reinforcement architectures are required, such as in the case of pressure vessels, aircraft fuselages and pipes. The process involves six main steps:

- Preparation and positioning of the mandrel on the machine.
- Definition of a programme for the part, specifying the winding pattern. The rotation of the mandrel is combined with the horizontal translation of a carriage guiding the reinforcement deposition, allowing the fibres to lay down in the desired pattern.
- Reinforcement impregnation, achieved by forcing the fibres to pass through a catalysed-resin bath. The excess resin is usually removed by means of a wiping device.
- Reinforcement deposition.
- Composite cure, at room temperature or in an autoclave.
- Mandrel ejection.

More detailed descriptions of the process are reported in [23, 24].

3.1.6 Compression moulding

The terms 'compression moulding' and 'press moulding' are used to indicate a class of manufacturing processes in which the shaping (or forming) of the part is achieved by means of the closure of two halves of a die mounted on a press [25, 26]. The induced compression forces the raw material to fill the cavity, providing good dimensional precision and surface finish at high production rates. Consolidation of the part is subsequently obtained by heating the die, in the case of thermoset processing. Compression moulding processes are commonly employed also to manufacture thermoplastic composites reinforced using short fibres. In this case, however, the system must be cooled more or less to room temperature in order to solidify the matrix.

The process is generally organized according to the following steps:

- Positioning of the appropriate amount of the raw material in the lower (possibly preheated) die. The raw material is usually in the form of cut sheet moulding compound (SMC), obtained by including continuous or chopped fibres and other fillers in a catalysed but uncured resin layer, or in the form of a bulk moulding compound (BMC).
- Die closure. The applied pressure forces the material to flow and fill the cavity, resulting also in the expulsion of air entrapped in the cavity and in the material. The moulding pressure varies over a broad range (1–35 MPa), depending on the initial material composition, the geometry of the part and the induced flow. Typical processing temperatures are in the range 130–160 °C.
- Material cure. The heat transferred from the die to the composite activates the exothermic reaction of the resin. The heating cycle should be planned to reduce the overall cycle time, at the same time avoiding excessive thermal peaks and chemical degradation of the resin system, in particular when thick parts are processed. This problem is commonly solved by increasing the percentage of filler in the raw material, decreasing the exothermic peak.
- Cooling and ejection of the product.

A further aspect of complexity arises when SMCs with long fibres (possibly woven) are processed, owing to displacement of the fibres, leading to geometrical defects such as wrinkling and modification of the reinforcement architecture.

3.2 Manufacturing of brake friction materials

3.2.1 Introduction

Friction materials are an important part of braking systems. They convert the kinetic energy of a car to thermal energy by friction in the contact zone. The tribological behaviour of brake lining materials plays a crucial role, where the complicated series of events that occur in the contact zone has to be taken into account [27]. The materials to be used for brake lining have to be developed to meet comprehensive tribological and mechanical requirements. They are a mix of more than 20 different ingredients. The friction material in an automotive brake system can be considered as one of the key components responsible for the overall functional characteristics of motor vehicle brakes. This is because it plays a key role in various aspects of brake performance, such as the stopping distance, pedal feel, wear and brake-induced vibration [28]. Furthermore, the friction materials for tomorrow have to be more comfortable and less noisy and still have low wear rates [29]. Competitive advantages have to be achieved in the brake friction materials industry by appropriate management of the formulation of friction materials and of changes in their manufacturing conditions, and the skilful implementation of these changes in a cost-effective manner. The management of these changes can have a decisive influence on whether a new friction material can be launched on the market. This is important because the friction material manufacturing industry has always been focused on greater customer satisfaction regarding (i) improved friction stability, (ii) improved life, (iii) absence of judder, (iv) absence of noise and (v) improved rotor compatibility [30].

The manufacture of brake friction materials is a complex process because powder mixes, which contain metal chips, filling agents, abrasives and phenolic resins, have to be moulded onto a backplate under pressure and temperature [31]. These moulded friction materials are then exposed to additional thermal treatments. This manufacturing process leads to anisotropic and viscoelastic behaviour of the material, as well as a certain amount of heterogeneity [31]. The moulding and heat treatment parameters of brake friction materials can often be chosen based on resin properties specified by the resin manufacturer. The manufacturing conditions of the friction materials for disc pads, for instance, are affected mostly by the thermal properties of the binder resin, such as the flow distance and the gel or 'B' time. Phenolic resins (modified and unmodified) are invariably used as binders in friction materials owing to their low cost and good combination of mechanical properties, such as high hardness, compressive strength, moderate thermal resistance, creep resistance and very good wetting capability with most ingredients [27]. The concentration of the phenolic resin in a friction material needs to be reconciled with the share of other ingredients in the formulation of the material. The primary use of phenolic resins is as a bonding agent. Phenolic resins easily penetrate and adhere to the structure of many organic and inorganic fillers and reinforcements, which makes them an ideal candidate for various end uses. The unique ability of phenolic resins to 'wet out' and cross-link throughout fillers and reinforcements provides a means to engineer the desired mechanical, thermal and chemical-resistance properties. The inherently heat-resistant phenolic resins carbonize and char at extreme service temperatures; they do not melt and smear like other polymer matrices. This property results in restored friction properties when the material cools and 'recovers' from hard braking. The formulas for phenolic composite friction materials are combinations of friction and wear-controlling agents, reinforcing fibres, and inert fillers blended with uncured phenolic resin in an amount necessary to bond the other ingredients in place with sufficient strength and resiliency when finished. The effect of the binder on the final composite's properties depends on the total formulation and manufacturing method. That is, no single type of resin product works optimally with all friction formulas or applications.

As has been mentioned, the important step in the manufacture of friction materials is the moulding and initial curing of the composite under heat and pressure. The moulding step typically involves pressing a uniform blend of ingredients in a shaped mould preheated to 150–200 °C under a pressure between 30 and 70 MPa. The phenolic resin melts and flows to coat the other ingredients and then secure them when the resin cross-links or 'cures' to an infusible state. The resin's performance during the hot moulding step is the most important aspect of assuring an efficient manufacturing process. Accordingly, manufacturing parameters such as the moulding pressure, moulding time and moulding temperature need to be properly chosen to be suitable for the formulation of the material. Decisions made at this stage strongly affect the future friction and wear behaviour of brake friction materials. It should be mentioned that there is a difference between the formulations for the original-equipment market and for the aftermarket. Differences between the performance of brake friction materials for the original-equipment market and the aftermarket are allowed by Regulation 90 of the Economic Commission for Europe of the United Nations [32, 33]. The formulation of a brake friction material addresses not only the selection of ingredients but also how they are dispersed in the mixture and the consistency of the final product [34]. On the other hand, different processing of the same formula for a friction material could change the porosity, compressibility, elasticity etc. That is why it is very important to know the effects of different brake friction material formulations in synergy with manufacturing conditions.

3.2.2 Manufacturing processes

The performance of a brake is primarily influenced by the contact situation between the cast iron brake disc and the friction material of the disc pads. However, the contact situation is affected in different ways by the wide diversity in the mechanical properties of the ingredients of the friction material [35]. These ingredients can be grouped into at least four classes: (i) fibrous reinforcements, (ii) binders, (iii) fillers and (iv) friction modifiers. Fillers are added to reduce cost or to impart desired performance. Fillers are used to lubricate (using graphite) the interface between the friction material and the brake disc or to increase the friction level (using abrasive). The diversity of the chemical and mechanical properties of the ingredients is changed further, in some extent, by the manufacturing parameters. This situation is additionally complicated if synergistic influences between these parameters are taken into consideration. The synergetic effects of all influencing factors, as shown in Figure 3.5, determine the final characteristics of a friction material. The basic steps in the disc pad manufacturing process consist of the following: (i) dry mixing, (ii) preforming (pre-moulding), (iii) moulding (press cure) and (iv) post-curing (see Figure 3.5). However, there are three stages that are considered as the most important ones that can affect the performance and physical properties of a brake friction material: (a) mixing, (b) preforming and (c) press cure.

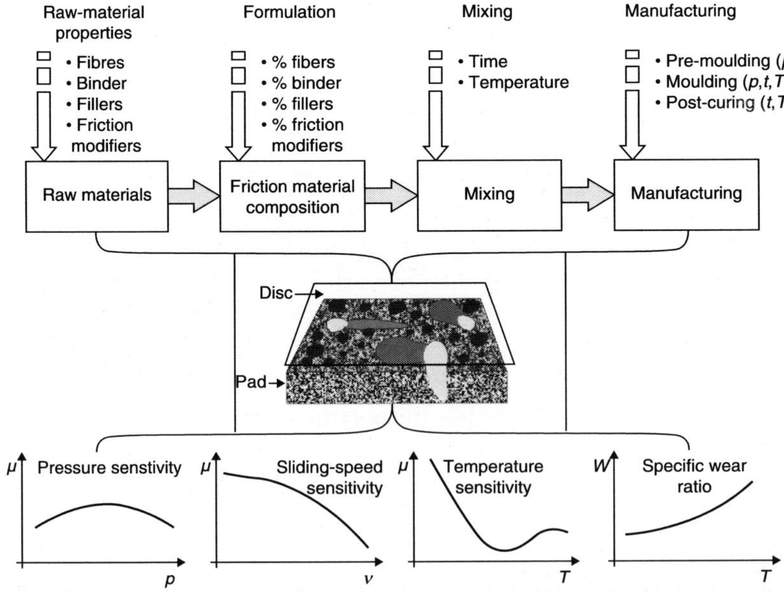

Figure 3.5 **Synergetic effects of formulation and manufacturing conditions on friction and wear of a brake friction material**

It is worth mentioning that dry mix materials were developed when asbestos fibre was used. Mixing cycles can vary, but total cycle times of between five and fifteen minutes are generally adopted [34]. This phase in the friction material manufacturing process is very important because high-speed mixing can produce shearing and chopping of the ingredients. The effectiveness of this phase in the preparation of the friction material is dependent upon the ploughs and choppers used [34]. The type of mixer and the mixing time chosen depend on the friction material formulation, i.e. the ingredients selected. In the case of dry mixes containing steel wool, excessive chopper action can reduce the particle size and change the shape of ingredients such as sponge iron powder. Aramid fibres need to be well opened, and special ploughs and dispenser blades have been recommended [34]. Glass fibres are sensitive to intensive mixing and prolonged mixing times. Preforming (pre-moulding) is preferred for dry-mixing materials. According to [34], friction materials have been bulky in the past. Curing moulds were designed to accommodate a compound with a high compression ratio (7: 1). Pre-compressing into compact preforms with a compression ratio of 2: 1 or 3: 1 permits simpler and lower-cost curing moulds. It allows flash moulds to be used for press curing, which is less expensive than positive curing moulds with devices for ejection [34].

From Figure 3.5, it can be concluded that the most complicated task is related to establishing a relationship between the formulation of the friction material and its manufacturing parameters versus the imposed requirements. This is particularly difficult considering the complexity of the requirements generally imposed on friction materials, relating to (i) the sensitivity of the friction coefficient to changes in the brake pressure actuation, the sliding speed and/or the brake interface temperature; (ii) the wear characteristics in different regimes of operating temperature; (iii) the vehicle weight; and (iv) the characteristics of the braking system [35]. Up to now, the development of new friction materials has mostly been done empirically [36], because very little is known about the interrelated influences in the contact zone during braking. Furthermore, third-body layers, which develop under different working conditions, particularly at elevated temperature, have increased the complexity of the brake contact situation. As explained so far and illustrated in Figure 3.5, there is no doubt that the optimal conditions for the development of a brake friction material are extremely difficult to achieve, and often soft computing techniques have to be employed in order to solve this complex problem [37–39].

According to [27], friction material moulding involves several stages: initial moulding, elastic–plastic deformation, and particle fracture or fragmentation. The initial stage is affected mostly by the particle size and shape. As the applied pressure is increased further, plastic deformation occurs locally at the interparticle contact points. The mechanical properties and quality of the particles are important factors, because they control the compressibility behaviour of the powder [27]. The consequences of the different stages in the manufacture of friction materials and their interrelated influences are illustrated in Figure 3.5 [35]. There are many ways to press-cure disc brake pads, for instance, from a completely manual flash mould method to automatic positive moulding (see Figures 3.6 and 3.7). Flash moulding is a moulding process which utilizes a constant volume for each part and where the excess material is 'flashed' outside the mould cavity. On the other hand, automatic shuttle-loading and unloading devices have dictated down-

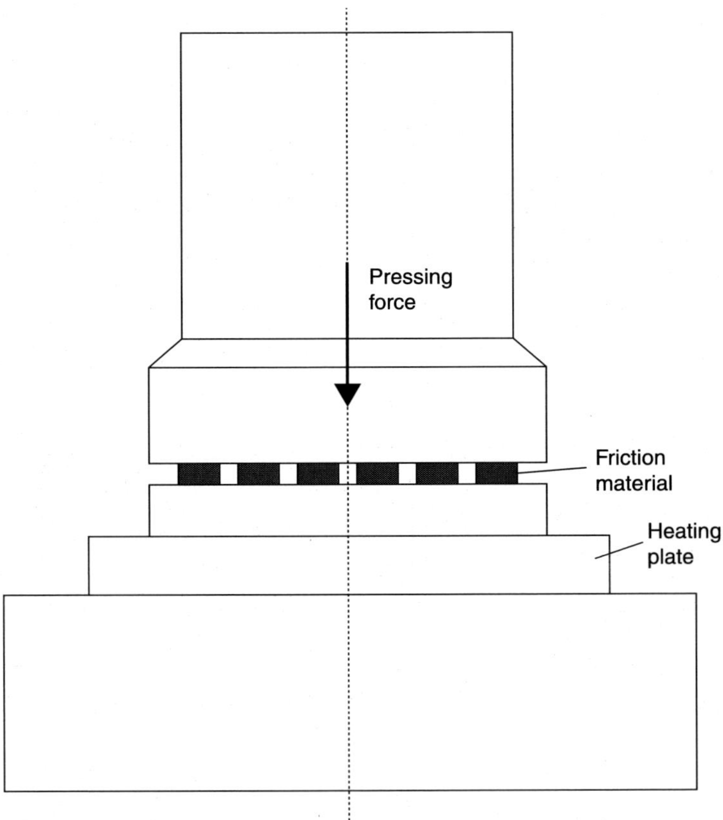

Figure 3.6 A flash mould method for brake friction material manufacturing

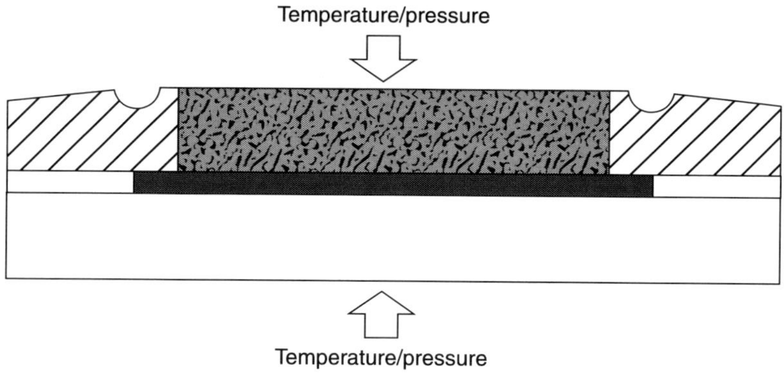

Figure 3.7 **Flash mould method – a mould cavity**

stroking presses to facilitate and improve the location of the tool in the press [34]. During so-called 'positive mould' press curing, the premixed ingredients are placed directly inside the cavity of the cure tooling. Full pressure is exerted on each part by a punch. This guarantees that full density is reached in every part. The composition of the friction material is not compromised by the process and can be tailored for the exact characteristics needed on a vehicle. In that case, automatic weighing has to be done precisely because any variations in weight from one cavity to another cause specific-gravity variations and changes in the mechanical characteristics of the friction material. In the 'flash mould' process, the resin content of the material must be increased to make the material flow. A high resin content creates fade, which could increase the stopping distance.

For instance, an important factor in the selection and development of brake friction materials is the compressibility value. This is critical to the quality aspects of a brake pad and may be influenced both by the manufacturing process and by the formulation of the friction material. The compressibility value is a measure of the pad deformation due to a standardized collection of loads. The compressive strain of a brake lining is an important design parameter in the evaluation of brake-fluid volume displacement, brake-pedal travel, and the propensity for judder or noise. The total pad deformation after a compressibility test is a combination of elastic and plastic deformation, as well as a creep portion depending on the viscosity of the material. In most passenger cars there is 10 to 20 mm of brake pedal travel before the brake torque builds up and vehicle deceleration begins [40]. Once contact is established, the deformation of the pads contributes to the pedal feel. There are several different compressibility test procedures (ISO6310, SAEJ2468 etc.).

The compressibility of a brake pad friction material should be within certain limits. It should ensure good pedal feel, minimal wear and sufficient power transfer from the pads to the disc. At the same time, the material should be elastic enough to ensure sufficient adhesion on

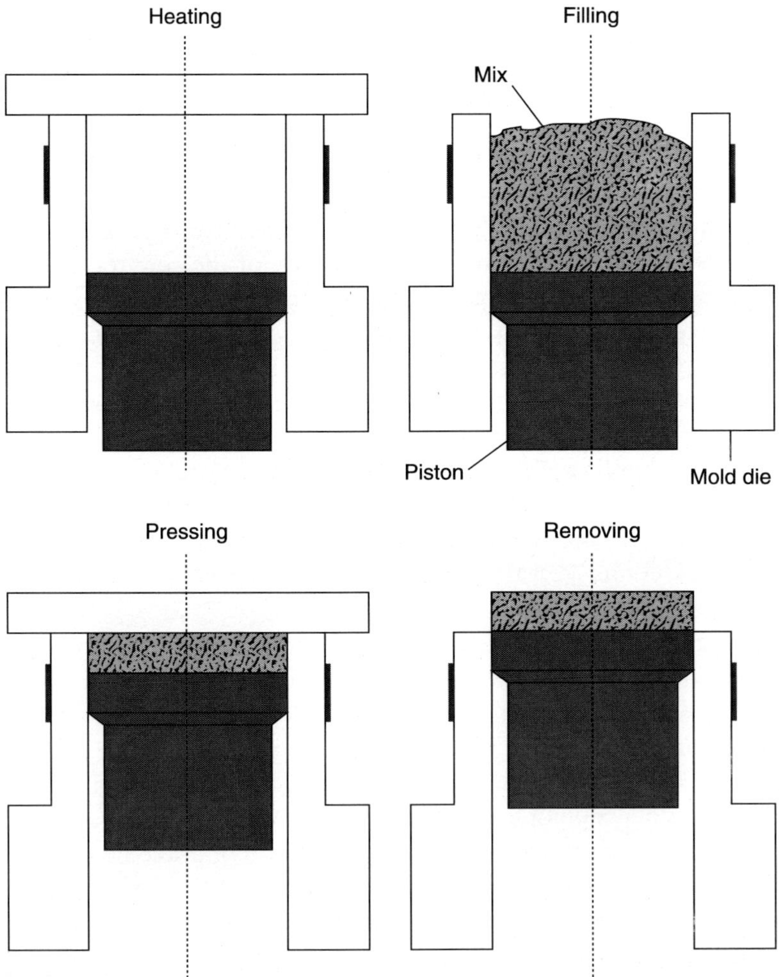

Figure 3.8 Positive moulding – compression process without a breathing cycle [5]

different disc surfaces. This makes it the most important indicator of manufacturing process quality, with respect to:

- the quality of the raw ingredients in the friction material mix;
- the way the friction material has been mixed;
- the quality and consistency of the pressing process;
- blistering during the press-curing process;
- cracks in the friction material;
- the bonding process between the friction material and the backplate.

As shown in Figure 3.8, the tool is heated to the required temperature with the piston in the lower part of the die [31]. The heating plate is removed and a predefined volume of the friction material mix is added from above. Then the heating plate is replaced on top of the die. The piston travels up and compresses the powder under pressure-controlled conditions. The compression process is interrupted by breathing cycles, to expel the gases produced in the cavity. Finally, the heating plate is again removed and the moulded part is ejected by the piston [31]. Since the compression phase is characterized by an increase in the piston force in line with the piston travel, the porosity of the mix is decreased and the particles are subjected to elastic and plastic deformation. It is important to select the optimum compression pressure for each new mix. According to [31], the friction material has a higher density on the piston side, and the density decreases towards the backplate side owing to frictional losses at the wall during compression.

3.3 Conclusions

The influence of manufacturing conditions on the characteristics of composite materials is extremely important. This is especially important in relation to the mechanical properties and the friction and/or wear performance of composite materials. Owing to the complex interrelated effects of the formulation of a composite material and the manufacturing parameters used, methods for predicting the effects of manufacturing parameters on the mechanical/friction/wear properties of the material need to be developed. Manufacturing conditions may cause substantial changes in the final properties of the composite material. The choice of the optimal manufacturing conditions needs to be made taking the formulation of the material and the desired properties into consideration.

It is obvious that improving the characteristics of composite materials requires the optimization of several different influencing factors. This optimization should be done especially in relation to the moulding process, which is the most important influencing factor on the performance of composite materials and their mechanical characteristics.

3.4 References

[1] Gutowski T.G. *Advanced Composites Manufacturing*, John Wiley & Sons, New York (1997).
[2] Miller R.R., Newhook R., Poole A. Styrene production, use and human exposure, *Critical Reviews in Toxicology*, 24, 1–10 (1994).
[3] Campbell F.C. *Manufacturing Processes for Advanced Composites*, Elsevier Advanced Technology, Oxford (2004).
[4] Mazumdar S.K. *Composites Manufacturing: Materials, Product, and Process Engineering*, CRC Press, Boca Raton (2000).
[5] Brent Strong A. *Fundamentals of Composites Manufacturing: Materials, Methods, and Applications*, Society of Manufacturing Engineers (2007).
[6] Mallick P.K. *Fiber-Reinforced Composites: Materials, Manufacturing, and Design*, CRC Press, Boca Raton (2007).
[7] Astrom P.T. *Manufacturing of Polymer Composites*, CRC Press, Boca Raton (1997).
[8] Oliver P., Cavarero M. Comparison between longitudinal tensile characteristics of thin and thick thermoset composite laminates: influence of curing conditions, *Computers and Structures*, 76(1–3), 125–37 (2000).
[9] Fernlund G., Rahman N., Courdji R., Bresslauer M., Poursartip A. et al. Experimental and numerical study of the effect of cure cycle, tool surface, geometry, and lay-up on the dimensional fidelity of autoclave-processed composite parts, *Composites Part A: Applied Science and Manufacturing*, 33(3), 341–51 (2002).
[10] Ruiz E., Trochu, F. Numerical analysis of cure temperature and internal stresses in thin and thick RTM parts, *Composites Part A: Applied Science and Manufacturing*, 36(6), 806–26 (2005).
[11] Rudd C.D., Long A.C., Kendall K.N., Mangin C.G.E. *Liquid Moulding Technologies: Resin Transfer Moulding, Structural Reaction Injection Moulding and Related Processing Techniques*, Woodhead (1997).
[12] Acheson J.A., Simacek P., Advani S.G. The implication of fiber compaction and saturation on fully coupled VARTM process, *Composites: Part A: Applied Science and Manufacturing*, 35, 159–69 (2004).
[13] Williams C., Summerscales J., Grove J. Resin infusion under flexible tooling (RIFT): a review, *Composites: Part A: Applied Science and Manufacturing*, 27, 517–24 (1996).

[14] Thagard J.R., Okoli O.I., Liang Z., Wang H.-P., Zhang C. Resin infusion between double flexible tooling: prototype development, *Composites: Part A: Applied Science and Manufacturing*, 34, 803–11 (2003).

[15] Han K., Jiang S., Zhang C., Wang B. Flow modelling and simulation of SCRIMP for composites manufacturing, *Composites Part A: Applied Science and Manufacturing*, 31, 79–86 (2000).

[16] Alms J., Advani S.G. Simulation and experimental validation of flow flooding chamber method of resin delivery in liquid composite molding, *Composites Part A: Applied Science and Manufacturing*, 38, 2131–41 (2007).

[17] Qi B., Raju J., Kruckenberg T., Stanning R. A resin film infusion process for manufacture of advanced composite structures, *Composite Structures*, 47, 471–6 (1999).

[18] Song Y.S., Youn J.R., Gutowski T.G. Life cycle energy analysis of fiber-reinforced composites, *Composites Part A: Applied Science and Manufacturing*, 40, 1257–65 (2009).

[19] Starr T.F. *Pultrusion for Engineers*, Woodhead (2000).

[20] Baran I., Tutum C.C., Nielsen M.W., Hattel J.H. Process induced residual stresses and distortions in pultrusion, *Composites Part B: Engineering*, 51, 148–61 (2013).

[21] Haffner S.M., Friedrich K., Hogg P.J., Busfield J.J.C. Finite-element-assisted modelling of a thermoplastic pultrusion process for powder impregnated yarn, *Composites Science and Technology*, 58, 1371–80 (1998).

[22] Kim D.H., Lee W.I., Friedrich K. A model for a thermoplastic pultrusion process using commingled yarns, *Composites Science and Technology*, 61, 1065–77 (2001).

[23] Rosato D.V., Grove C.S. *Filament Winding: Its Development, Manufacture, Applications and Design*, John Wiley & Sons (1964).

[24] Peters S.T. *Composite Filament Winding*, ASM International, Materials Park, OH (2011).

[25] Isayev, A.L. *Injection and Compression Molding Fundamentals*, CRC Press, Boca Raton (1987).

[26] Davis B.A., Gramann P.J., Osswald T.A., Rios A.C. *Compression Molding*, Carl Hanser, Munich (2003).

[27] Rukiye E., Nurettin Y. An experimental study on the effects of manufacturing parameters on the tribological properties of brake lining materials, *Wear*, 268, 1524–32 (2010).

[28] Kima Y.C., Cho M.H., Kim S.J., Jang H. The effect of phenolic resin, potassium titanate, and CNSL on the tribological properties of brake friction materials, *Wear*, 264, 204–10 (2008).

[29] Van Berchum B., Borsboom F., Verheijen M., Husson M. Development of a rubber coated mineral fiber for disc pad applications, SAE Technical Paper 2002-01-2596 (2002), doi:10.4271/2002-01-2596.

[30] Sasaki Y. Development philosophy of friction materials for automotive disc brakes, *Proceedings of the Japan Society of Automotive Engineers*, No. 9531679, 407–12 (1995).

[31] Schiffner, K., Heftrich, M., Brecht, J. Modeling of compaction processes of friction material mixes, SAE Technical Paper 2002-01-2594 (2002), doi:10.4271/2002-01-2594.

[32] Economic Commission for Europe of the United Nations, Regulation No. 13-H – Uniform provisions concerning the approval of passenger cars with regard to braking (2012).

[33] Economic Commission for Europe of the United Nations, Regulation No. 90 – Uniform provisions concerning the approval of replacement brake lining assemblies, drum brake linings and discs and drums for power-driven vehicles and their trailers (2012).

[34] Nicholson G. *Facts about Friction*, Gedoran (1995).

[35] Aleksendrić D., Duboka Č. Automotive friction material development by neural computation, *Braking 2006*, York, United Kingdom, 167–76 (2006).

[36] Decuzzi P., Demelio G. The effect of material properties on the thermoelastic stability of sliding systems, *Wear*, 252, 311–21 (2002).

[37] Aleksendrić D., Duboka Č. Prediction of automotive friction material characteristics using artificial neural networks – cold performance, *Wear*, 261, 269–82 (2006).

[38] Aleksendrić D., Duboka Č. Fade performance prediction of automotive friction materials by means of artificial neural networks, *Wear*, 262, 778–90 (2007).

[39] Aleksendrić D. Artificial neural networks in automotive brakes' friction material development, Doctoral thesis, Faculty of Mechanical Engineering, University of Belgrade (2007).

[40] Basch R., Sanders P., Hartsock D., Evans C. Correlation of lining properties with brake pedal feel, SAE Technical Paper 2002-01-2602 (2002), doi:10.4271/2002-01-2602.

Soft computing techniques

DOI: 10.1533/9781782421801.39

Abstract: In this chapter, the basic characteristics of artificial neural networks and genetic algorithms are presented. Artificial neural networks and genetic algorithms, as techniques of artificial intelligence, demonstrate powerful problem-solving ability. They are based on quite simple principles, but take advantage of their mathematical nature based on non-linear iteration. The goal of an artificial neural network is to learn or to discover some association between input and output patterns, in order to analyse or find structure in the input patterns. Recent research has shown that remarkable benefits can be achieved by applying these techniques in conjunction with experimental data or in combination with other computational methods, also described in this chapter. Genetic algorithms represent search algorithms based on the mechanics of natural selection and natural genetics. They have been used for many optimization tasks, in particular by combining them with other techniques of artificial intelligence.

Key words: artificial neural networks, genetic algorithms, computational methods, hybrid models.

4.1 Artificial neural networks

Inspired by the biological nervous system, artificial neural network (ANN) technology is currently being used to solve a wide variety of complex scientific, engineering and business problems. ANNs are ideally suited to such problems because, like their biological counterparts, they can learn and therefore can be trained to find solutions, recognize patterns, classify data and forecast future events [1]. An implementation of artificial intelligence should provide development of optimal design using methods that incorporate adaptation and learning [2, 3]. According to [4] and Figure 4.1, an ANN is a system composed of many simple processing elements operating in parallel, whose function is determined by the network structure, the connection strengths, and the processing performed at computing elements or nodes. A neural network is a massively parallel distributed processor that has a natural propensity for storing experiential knowledge and making it available for use. It resembles the brain in two respects: (i) knowledge is acquired by the network through a learning process, and (ii) the interneuron connection strengths, known as synaptic weights, are used to store this knowledge [4].

That is why ANNs can be most adequately characterized as 'computational models' with particular properties, such as the ability to adapt or learn, to generalize, and to cluster or organize data. The important benefit provided by the operation of ANNs is their parallel data processing. The main distinction between the types of ANNs can be made as follows: (i) feedforward networks, where the data flow from the input to the output units is strictly feedforward, and (ii) recurrent networks, which do contain feedback connections. In contrast to feedforward networks, the dynamical properties of the network are important in the latter case. In some cases, the activation values of the units undergo a relaxation process such that the network evolves to a stable state in which these activations do not change any more. In other applications, the changes in the activation

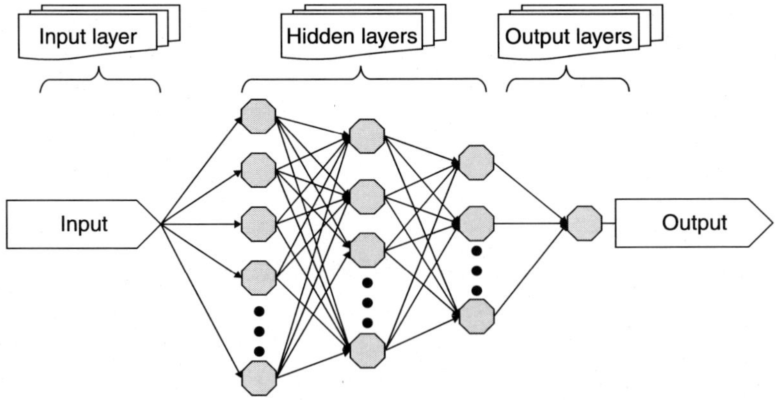

Figure 4.1 The basic architecture of an artificial neural network

values of the output neurons are significant, such that the dynamical behaviour constitutes the output of the network.

The goal of an ANN is to learn or discover some association between the input and output patterns, and to analyse or find the structure of the input patterns. The learning process is achieved through modification of the weights of the connections between units (see Figure 4.2). The learning process of neural networks can be categorized into two distinct sorts. One of these is supervised learning or associative learning, in which the network is trained by providing it with input patterns and matching output patterns. These input–output pairs can be provided by an external teacher, or by the system which contains the network (self-supervised). On the other hand, there is also unsupervised learning or self-organization, in which an (output) unit is trained to respond to clusters of patterns within the input. In this paradigm, the system is expected to discover statistically salient features of the input population.

Basically, neural networks are built from simple units, sometimes called neurons or cells by analogy with their real counterparts. These units are linked by a set of weighted connections. Learning is usually accomplished by modification of the connection weights. Each unit codes or corresponds to a feature or characteristic of a pattern that we want to analyse or that we want to use as a predictor. Artificial neural networks are adaptive statistical devices. This means that they can change the values of their parameters (i.e. the synaptic weights) iteratively as a function of their performance. These changes are made according to learning rules, which can be characterized as supervised (when the desired

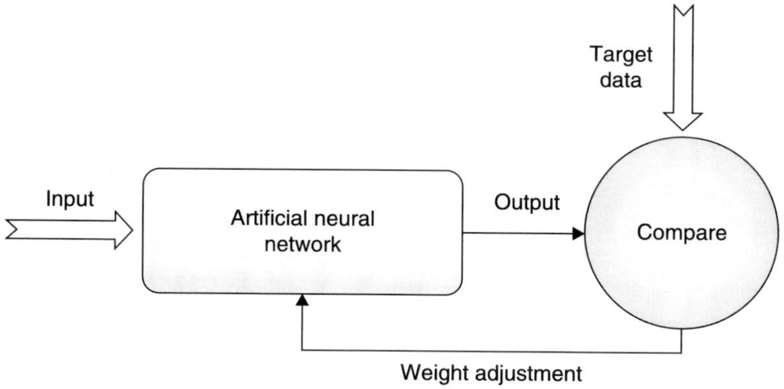

Figure 4.2 **The learning process of an artificial neural network**

output is known and is used to compute an error signal) or unsupervised (when no such error signal is used) [5]. A feedforward network has a layered structure, as depicted in Figure 4.1. Each layer consists of units which receive their input from units in a layer directly below and send their output to units in a layer directly above that unit.

The network processes a number of inputs from the outside world to produce an output (i.e. the network's predictions). The neurons are connected by weights and grouped into layers by their association with the outside world. For example, if a neuron receives data from outside of the network, it is considered to be in the input layer. If a neuron contains the network's predictions, it is in the output layer. Neurons between the input and output layers are in the hidden layer, which serve to (i) add non-linearity to the system and (ii) address interactions between input variables. There can be many hidden layers, i.e. many levels of non-linearity and many interactions [6]. The number of hidden-layer neurons is one of the most important considerations when solving problems [7].

The description of the architecture of a network specifies how many layers the network has, the number of neurons in each layer, the transfer function of each layer and how the layers are connected to each other. Defining an ANN architecture, according to [8, 9], means determining the numbers of input, output and hidden neurons, and the number of hidden layers. In order to resolve this question, an equation relating the numbers of units in the input, hidden and output layers has been proposed [10]. According to [4, 10], the major issue is to determine the number of hidden units and the weight coefficients. Various methods to set the strengths or weights of the connections between neurons exist. One way is to set the weights explicitly using a priori knowledge; another way is to train the neural network by feeding it teaching patterns. The initially set weights are then changed according to a predefined learning rule.

The best architecture to be used depends on the type of problem to be represented by the network. A network performs a computation by mapping input values to output values. That is why the ANN technology requires three sets of data, which are the training data, the validation data and the test data. The training data set is used to guide the training process, i.e. to update the weights of the neural network during training. The number of training patterns influences the generalization capacity of the network. That number, according to [8, 9], should be at least several times larger than the capacity of the ANN. The validation data set is used to monitor the quality of the training of the neural network model, to indicate when to terminate the training process. The test data set is used to examine the final quality of the neural model developed. The error signal is the product of the differences between the desired and actual output values. This error signal is then back-propagated to the lower layers in order to find the global minimum of the error surface area. It is obvious that neural-network training requires a sufficient number of input and output pairs in order to approximate the relationship between them, and not only for the data which are included in the training set [4, 11]. The process of development of an ANN model is shown in Figure 4.3.

Artificial neural networks have been shown to be an effective method for the prediction of time series events. Neural networks can be classified into dynamic and static categories. Static (feedforward) networks have no feedback elements and contain no delays – the output is calculated directly from the input through feedforward connections [10]. Dynamic neural networks are generally more powerful than static networks since they have a sort of memory that can remember the past values and states of the network [12, 13]. The output of a dynamic network depends on the current input values as well as on the previous inputs, outputs or states of the network. For this reason, the gradient must be computed in a more complex way. Dynamic neural networks can be trained using the same learning algorithms that are used for static networks. Traditional research in this area uses a network with a sequential iterative learning process based on the feedforward backpropagation approach [12]. In general, dynamic neural networks can be divided into two types: (1) feedforward time-delay networks, and (2) feedback or recurrent

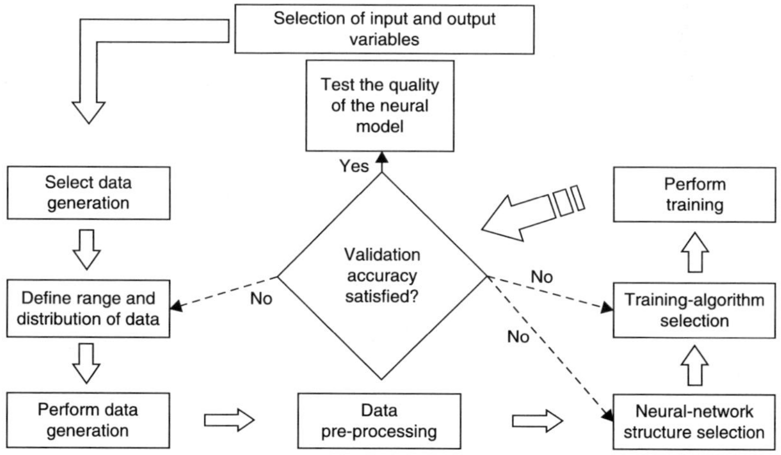

Figure 4.3 **The process of development of an artificial neural network model**

networks. Recurrent neural networks are built in such a way that the outputs of some neurons are fed back to the same neurons or to neurons in the preceding layers. They have an intrinsic dynamic memory: their outputs, at a given instant, reflect the current input, as well as the previous inputs and outputs, which are gradually quenched. The network function is largely determined by the interconnections between the neurons (the connection weights) [14]. According to [13, 15–16], the output of a dynamic network is affected differently by the weights. Two different effects can be considered: (i) a direct effect (a change in a weight causes an immediate change in the output at the current time step), and (ii) an indirect effect (which implies using dynamic backpropagation to compute the gradients, which is computationally more intensive). As a subcategory of dynamic neural networks, a layer-recurrent network is shown in Figure 4.4. This recurrent neural network is characterized by a feedback loop with a single delay around each layer of the network except for the last (output) layer (see Figure 4.5). It has found application in many filtering and modelling tasks.

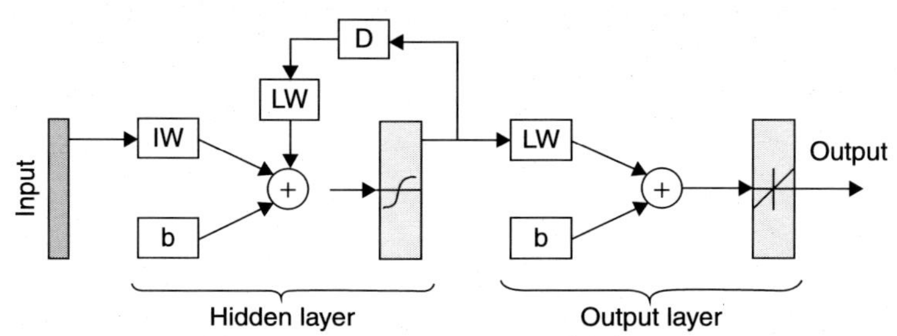

Figure 4.4 Typical structure of a layer-recurrent neural network [15, 16]

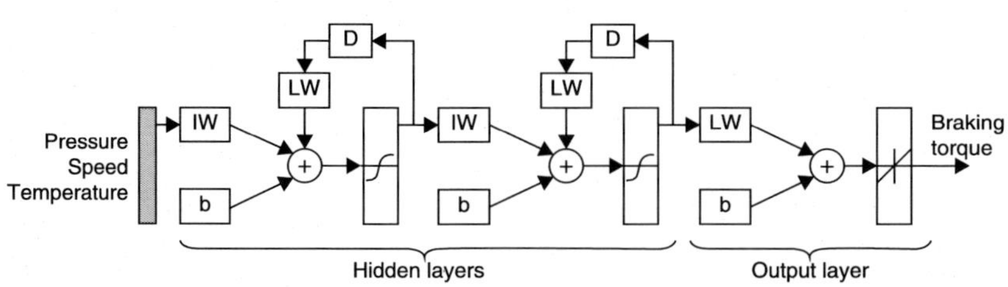

Figure 4.5 Dynamic neural model of disc brake operation based on a layer-recurrent network [15, 16]

Because neural network models show excellent approximation abilities, they are recognized as being able to approximate the non-linear behaviour of several dynamic processes with reasonable accuracy [15, 16]. A very important feature of these networks is their adaptive nature, where 'learning by example' replaces 'programming' in solving problems [15, 16]. They can be trained to perform a particular function by adjusting the values of the weight factors between the neurons, either by considering information coming from outside the network or by means of the neurons themselves in response to the input. The learning and recognition of patterns in large data sets is the key ability of neural networks for achieving learning and memory. It is very important in the case of the development of dynamic models. A dynamic neural model of the operation of a disc brake, for instance (see Figure 4.5), should recognize the dynamic correlation between changes in the input and output. For this reason, the input parameters for such a dynamic model are defined by the operation conditions of the disc brake (speed, pressure and brake interface temperature). A synergy of these influences affects the changes in the braking torque. The pattern of influence of the brake actuation pressure needs to be recognized versus the dynamic changes in the other braking regimes and the braking torque. That is why the braking torque is taken as the output parameter which represents the brake performance (Figure 4.5). As can be seen, the influence of previous values of the braking torque on its current value is represented through feedback connections after each of the hidden layers, known as layer delay states.

4.2 Artificial neural networks and other computational methods (FEM, FVM, FDM)

As aforementioned, the reliability of an artificial neural network, independently of its specific nature, is limited to the space covered by the initial data set and relies strongly on the amount of data available for training. In most cases, training of the network is performed using experimental data provided by appropriately designed experiments. It should be noted, however, that extensive experimental campaigns and the subsequent post-processing of results are generally seen as excessively expensive in time and cost, and sometimes as prohibitive. The relevance of these factors increases with the extension of the space covered by the design of experiments and the density of test cases, taking into account also that appropriate repetition of each test is desirable in order to ensure statistical validity of the analysis.

As an illustrative example, let us consider a very common process for thermoset composite manufacturing, namely the autoclave curing process described in Chapter 3. The tasks connected with this process generally start with the preparation of the material, such as the matrix, accelerators, catalysers and reinforcing layers, including also other tools, and their subsequent deposition and bagging. These activities can be accomplished by means of different techniques that are potentially highly labour-intensive. Differently from other traditional materials (e.g. aluminium or steel) and conventional manufacturing processes (e.g. turning or milling), the impact of the raw materials on the overall cost must be considered accurately. What is more, if the focus of the analysis is strictly related to the evolution of the manufacturing process itself (for instance, the evolution of the temperature or the degree of cure), the workpiece must be properly equipped with dedicated sensors for data acquisition. After this preliminary preparation, the material is placed appropriately into the autoclave and then subjected to appropriately planned temperature, pressure and vacuum cycles, which in many cases are highly energy-demanding. The need to activate and promote the exothermic resin reaction, avoiding any excessive temperature peaks or gradients and inducing, at the same time, inside–out solidification, imposes the adoption of cycles characterized by slow heating ramps and relatively long holding steps. In this way, a satisfactory degree-of-cure distribution and reduced residual stresses and distortions in the processed part are usually achieved; however, the time of residence in the autoclave can reach several hours (depending on the resin system and workpiece dimensions). At the end of the process, the manufactured product needs to be subjected to specific inspections and tests to evaluate the features of interest (degree of cure, percentage of voids, and residual stresses and distortions, for instance). Even assuming that all of the samples required for the statistical analysis could be manufactured in a single process, the need to explore different curing cycles could require weeks (or even months) and a large amount of money before the beginning of network training.

In this regard, intriguing opportunities exist to combine soft computing techniques (i.e. ANNs) with more 'traditional' numerical methods, based on the development of reliable theoretical models of the process considered and their subsequent solution by iterative computational approaches. Similarly to ANNs, the credibility of these computational models needs to be ensured by appropriate verification (the process of determining that the implementation of a model accurately represents the developer's conceptual description of the model and its solution) and validation (the process of determining the degree to which a model is an accurate representation of the real world from the perspective of the intended uses of the model). As far as the numerical modelling of thermoset-composite manufacturing processes is concerned, a vast literature is currently available, as outlined in the next chapter. In the interests of completeness, the basics of the most commonly adopted numerical methods, i.e. the finite difference method (FDM), the finite element method (FEM) and the finite volume method (FVM), are briefly recalled here. Generally speaking, the aforementioned methods allow one to solve, or to approximate the solution of, a problem formulated as a system of partial differential equations representing the governing (balance or equilibrium) equations which describe the relevant phenomena in the space and time domain considered, according to prescribed (typically Dirichlet or Neumann) boundary conditions. The solution provided by the application of these methods should be considered as an approximation to the ideal solution, conditioned by discretization and truncation errors.

4.2.1 Finite difference method

The basic concept of the FDM is to replace, under an appropriate hypothesis, the derivatives defining the governing equations by means of finite differences or finite quotients. This idea was originally conceived by Leonhard Euler (1707–1783) and successively extended thanks to contributions by C. Runge (1856–1927), M.W. Kutta (1867–1944), J. Crank (1916–2006), P. Nicolson (1917–1968) and John von Neumann (1903–1957), among others. For detailed reading on this subject, see [17–20]; in what follows, the main concepts are briefly recalled.

Consider a generic function u, twice continuously differentiable in the neighbourhood of $\bar{x}$; by definition, its first derivative can be written as

$$\frac{\partial u}{\partial x}(\bar{x}) = \lim_{\Delta x \to 0} \frac{u(\bar{x}+h)-u(\bar{x})}{h}. \qquad [4.1]$$

Therefore, if h is sufficiently small, the quotient on the right-hand side of Eq. 4.1 represents an acceptable approximation of the derivative:

$$u'(\bar{x}) \approx \frac{u(\bar{x}+h)-u(\bar{x})}{h}. \qquad [4.2]$$

The error involved in this approximation can be estimated using the Taylor series expansion of the function u:

$$u(\bar{x}+h) = u(\bar{x}) + \frac{u'(\bar{x})}{1!}h + \frac{u''(\bar{x})}{2!}h^2 + \cdots + \frac{u^{(n)}(\bar{x})}{n!}h^n + R_n(\bar{x}), \qquad [4.3]$$

where $R_n(\bar{x})$ is a remainder term.

Limiting the expansion to the first two terms on the right-hand side of Eq. 4.3, after simple manipulations, it follows that

$$\frac{u(\bar{x}+h) - u(\bar{x})}{h} = u'(\bar{x}) + \frac{o(h^2)}{h}, \qquad [4.4]$$

and, for sufficiently small values of h, this results in

$$\left| \frac{u(\bar{x}+h) - u(\bar{x})}{h} - u'(\bar{x}) \right| \le Ch, \qquad [4.5]$$

where C is a positive constant. In other words, the error committed by the first-order (forward) approximation in Eq. 4.2 is of order h.

In a similar way, assuming the following definition of the first derivative of the function u,

$$\frac{\partial u}{\partial x}(\bar{x}) = \lim_{\Delta x \to 0} \frac{u(\bar{x}) - u(\bar{x}-h)}{h}, \qquad [4.6]$$

the following writing of the first-order backward finite difference approximation of $u'(\bar{x})$ is obtained:

$$u'(\bar{x}) \approx \frac{u(\bar{x}) - u(\bar{x}-h)}{h}. \qquad [4.7]$$

It is worth noting that the order of the approximation is conditioned by the regularity of the function considered. Indeed, if u is only twice continuously differentiable in the neighbourhood of $\bar{x}$, then the approximation is consistent at order one only. However, if u is three times continuously differentiable in the neighbourhood of $\bar{x}$, the following definition of the first derivative of u,

$$\frac{\partial u}{\partial x}(\bar{x}) = \lim_{\Delta x \to 0} \frac{u(\bar{x}+h) - u(\bar{x}-h)}{h}, \qquad [4.8]$$

yields

$$u'(\bar{x}) \approx \frac{u(\bar{x}+h) - u(\bar{x}-h)}{2h}, \qquad [4.9]$$

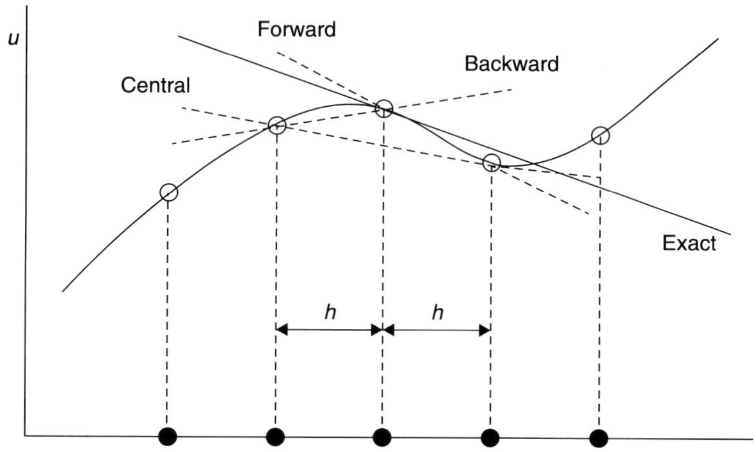

Figure 4.6 **Geometric interpretation of the finite difference approximation of the first derivative**

which represents the second-order central finite difference approximation of $u'(\bar{x})$. The geometric interpretation of the above equations is depicted in Figure 4.6.

Analogously, second-order derivatives can be written as follows:

$$u''(\overline{x}) \approx \frac{u(\overline{x}+h)-2u(\overline{x})+u(\overline{x}-h)}{h^2}, \qquad [4.10]$$

$$u''(\overline{x}) \approx \frac{u(\overline{x}+2h)-2u(\overline{x}+h)+u(\overline{x})}{h^2}, \qquad [4.11]$$

$$u''(\overline{x}) \approx \frac{u(\overline{x})-2u(\overline{x}-h)+u(\overline{x}-2h)}{h^2}, \qquad [4.12]$$

which represent the central, forward and backward first-order finite difference approximations, respectively, of $u''(\bar{x})$. As far as time derivatives are concerned, it should be noted that the direction of the discretization poses some stability and convergence issues. In particular, when a forward difference is used for the approximation of time derivatives, an explicit method is adopted, which is conditionally stable and convergent. On the other hand, an implicit method (the backward difference for approximation of time derivatives) is unconditionally stable and convergent, but more computationally expensive.

Intuitively, the computational domain is replaced by a grid, whose nodal points are connected by linear combinations of function values, providing a set of equations to be solved iteratively by means of one of the several numerical techniques available.

4.2.2 Finite element method

Similarly to the FDM, the FEM is included in a class of numerical techniques used to find a suitable approximation to the solution of a consistently formulated boundary value problem. Compared with the FDM, the FEM provides the remarkable advantage of being able to handle geometrically complex domains with relative simplicity. The basic idea of the FEM is to represent a continuous domain by geometrically simpler non-overlapping discrete entities, called finite elements or just elements, whose response is quantified by the values of degrees of freedom at the end (or nodal) points [21–24].

Early suggestions of the intriguing capabilities related to the application of this simple concept can be found in the work of A.R. Collar, W.J. Duncan, A. Hrennikoff and R. Courant, although the inventor of the FEM is generally recognized to be M.J. Turner, who generalized and improved the direct stiffness method. Great impetus was also given to the development of the FEM by the contributions of J.H. Argyris, H.C. Martin, R.W. Clough and O.C. Zienkiewicz, who transferred this technique from the aerospace industry to other engineering sectors (in particular, civil engineering). Other relevant contributions to the field are attributable to B.M. Irons, who conceptualized iso-parametric elements, numerical integration, shape functions and the patch test; R.J. Melosh, who demonstrated that conforming displacement models are a form of Rayleigh–Ritz method; E.L. Wilson, who developed the first open source FEM code; and, more recently, T.J.R. Hughes and K.J. Bathe. Nowadays, the dominant role of the FEM as a numerical analysis technique in engineering (in particular, in structural mechanics) is widely recognized, even if in some disciplines, such as computational fluid dynamics, the FDM or FVM is generally preferred.

The FEM belongs to the Galerkin residual methods, i.e. it is based on approximation of the solution of a generic problem by considering its weak formulation as a starting point. Briefly, the finite element approach to a generic boundary value problem is based on the subdivision (discretization) of the domain into subdomains (primitive elements), represented mathematically by equations descriptive of the problem. Elements can have an intrinsic dimensionality of zero, one, two or three space dimensions, as depicted in Figure 4.7.

The response of each element is expressed in terms of a finite number of degrees of freedom (rather than an infinite number as for a continuous domain) at nodal points and, in each element, the solution is given by a linear approximation of a shape (or basis) function. Typical basis

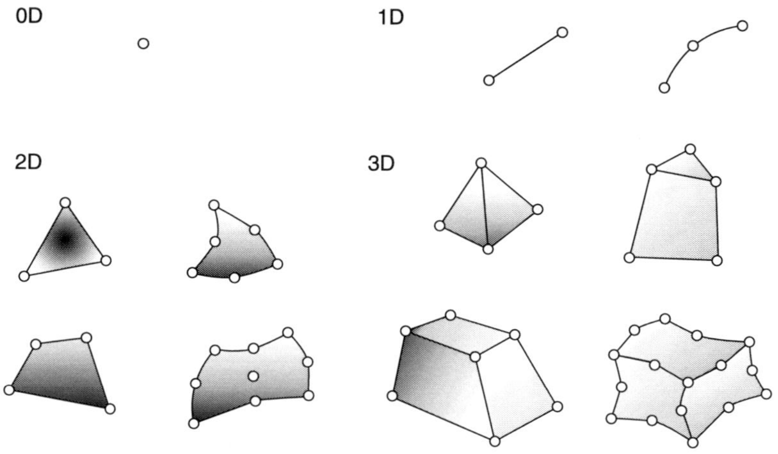

Figure 4.7 **Element type and dimensionality**

functions are linear or polynomial. Nodes provide the needed connections (and continuity) between adjacent elements and host the degrees of freedom.

The next step is the recombining (assembly) of the set of element equations in order to write a global system of equation, in the following form:

$$Ku = f, \qquad [4.13]$$

where K is the global stiffness (or conductivity) matrix, u is the vector of the unknown nodal degrees of freedom, and f is the vector of the applied load (or the boundary conditions). In other words, the stiffness (or conductivity) matrices of the elements and consistent nodal load vectors are merged, providing a master stiffness (or conductivity) matrix and a master load vector. Finally, a numerical answer to Eq. 4.13 is obtained by means of direct methods, which usually take advantage of the fact that finite element matrices are normally sparsely populated, or by means of iterative methods, such as the conjugate gradient method.

4.2.3 Finite volume method

The FVM is a numerical method used to evaluate elliptic, parabolic or hyperbolic partial differential equations in the form of algebraic equations, on the basis of conservation laws. Its development, attributable to the work of S. Patankar, R.J. Leverque, E.F. Toro and R. Eymard, among others, is relatively recent [25–28].

Similarly to the aforementioned techniques, the FVM evaluates the variables at discrete locations in the computational domain. Like the FEM, the FVM divides a geometrically arbitrary domain into a finite number of elements (a structured or unstructured mesh), subsequently used to build finite or control volumes (a dual mesh). The discretization of the domain into control volumes can be performed by adopting a vertex-centred approach (where each node of the mesh is the centre of a finite volume, whose boundaries are obtained by connecting the centroids of each element and the midpoints of each element edge), rather than a cell-centred approach (where control volumes coincide with elements), as depicted in Figure 4.8 for the two-dimensional case.

After the domain decomposition, an integral formulation of the balance equations and a successive approximation of integrals by numerical integration are needed for each control volume. Considering, for instance, the mass transport balance for an incompressible fluid, under the hypothesis of a stationary (non-deformable) mesh and neglecting source terms, its conservation law can be written

$$\frac{\partial \rho}{\partial t} + \nabla \cdot \rho v = 0, \qquad [4.14]$$

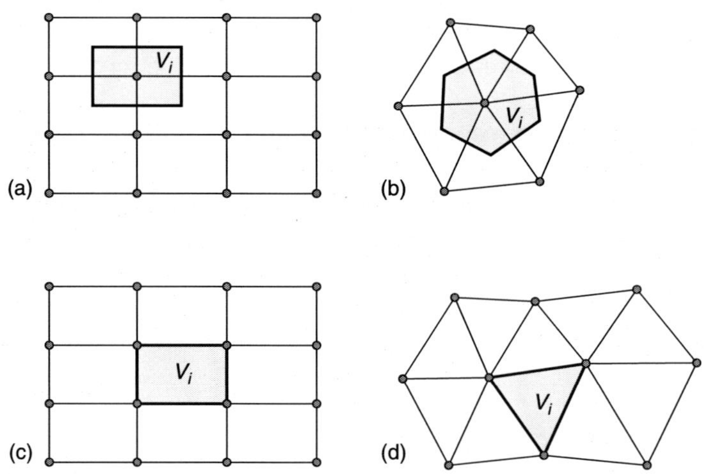

Figure 4.8 **Mesh and dual mesh in vertex-centred FVM (a, b) and cell-centred FVM (c, d). Control volumes are defined by the grey-coloured areas**

where ρ is the density and v is the velocity vector. The basic idea of the FVM is to enforce Eq. 4.14 for a small control volume defined by the computational mesh; therefore, in the ith cell, characterized by a volume V_i and boundary surface S_i, the result is

$$\int_{V_i} \frac{\partial \rho}{\partial t} dV + \int_{V_i} \nabla \cdot \rho v \, dV = 0. \qquad [4.15]$$

Since finite volume methods discretize the balance equation directly, an obvious virtue of such methods is the conservation property (the flux entering a given volume is identical to that leaving the adjacent volume), in comparison with the weak formulation adopted in the FEM. Because of this feature, the FVM has proved to be very suitable for the solution of problems in fluid mechanics, as well as in heat and mass transfer.

The integral conservation law in Eq. 4.15 can be rewritten by volume-averaging the first term and applying the Gauss theorem to the second term. Simple manipulations yield

$$\frac{d\rho_i}{dt} + \frac{1}{V_i} \oint_{S_i} \rho v \cdot n \, dS = 0, \qquad [4.16]$$

where n is the normal vector to S_i, outward from V_i.

Function values and derivatives can be approximated in a finite difference fashion. Adopting a first-order Euler scheme, this results in the following for the first term of Eq. 4.16:

$$\frac{d\rho_i}{dt} \approx \left(\frac{\rho - \rho^{(n-1)}}{\Delta t} \right), \qquad [4.17]$$

where Δt is the time step and the index $(n-1)$ refers to the previous temporal step. The second term can be treated as a summation extended to all integration points of the control volume. The equations written for each volume are then assembled, providing a unique algebraic system to be solved using numerical methods. Owing to the similarity with the FDM and FEM, the FVM is often referred to also as the control volume finite element method, the cell-centred finite difference method or the generalized finite difference method. The main difference between the finite volume scheme and the finite difference scheme lies in the fact that the finite difference approximation is used for the flux rather than for the operator itself. Furthermore, the FVM is more flexible than standard finite difference methods, which are mainly defined on structured grids, simple domains and homogeneous materials.

4.3 Genetic algorithms

Genetic algorithms (GAs) are a family of computational models inspired by evolution. These algorithms encode a potential solution to a specific problem on a simple chromosome-like data structure and apply some recombination operators to this structure so as to preserve critical information. GAs are often viewed as function optimizers, although the range of problems to which genetic algorithms have been applied is quite broad [29, 30]. Basically, GAs work with a 'population' of 'individuals', each representing a possible solution to a given problem. Each individual is assigned a 'fitness score' according to how good a solution to the problem it is. The highly fit individuals are given opportunities to 'reproduce' by 'cross-breeding' with other individuals in the population. This produces new individuals as 'offspring', which share some features taken from each 'parent'. The least fit members of the population are less likely to get selected for reproduction, and so 'die out' [29, 30].

Genetic algorithms represent search algorithms based on the mechanics of natural selection and natural genetics. They have been used for many optimization tasks, in particular by combining them with other techniques of artificial intelligence, as shown in [31–36]. One of the most fundamental principles in our world is the search for an optimal state [32]. The basic GA optimization cycle is shown in Figure 4.9. GAs start with an initial

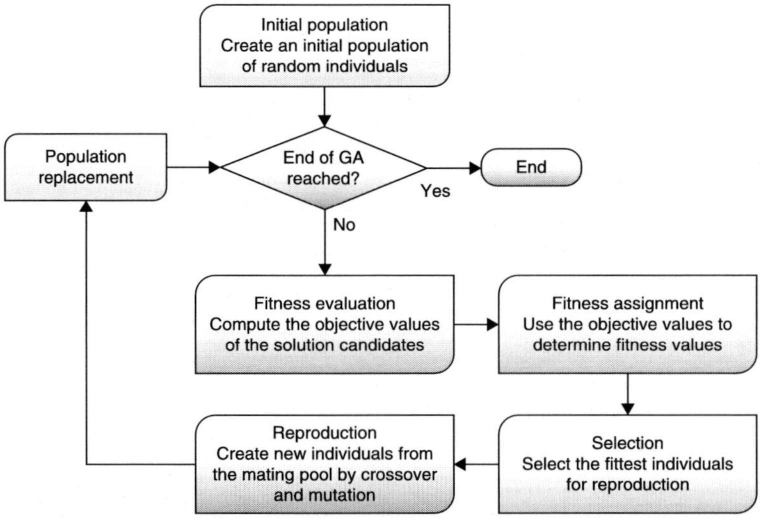

Figure 4.9 **The basic cycle of a genetic algorithm [12]**

population of individuals (chromosomes) randomly chosen from the design space and search the input range effectively for the required output variables by means of genetic operators, namely reproduction, crossover and mutation [37–40]. A population of chromosomes (individuals) represents a set of potential solutions to the problem to be solved. In many cases this consists in the optimization of a generally very complex function [40]. In every evolutionary step, known as a generation, the individuals in the current population are decoded (evaluated) according to some predefined quality criterion, referred to as the fitness function, i.e. the objective function [31]. This means that a GA repeatedly modifies a population of individual solutions. Therefore, each individual in the population has an associated fitness, indicating the utility or adaptation of the solution that it represents [40]. The most common representation of the chromosomes is binary. The chromosomes consist of a set of genes, which are generally characters belonging to the alphabet {0,1}. A GA uses three main types of rules at each step to create the next generation from the current population: (i) *selection*, to select the individuals, called parents, that contribute to the population at the next generation; (ii) *crossover*, to combine two parents to form children for the next generation; and (iii) *mutation*, to apply random changes to individual parents in order to induce some alterations in the children. At each step, the GA selects some individuals from the current population, following a probabilistic criterion, to become parents; in other words, these individuals are used to populate the next generation. Since crossovers between the best individuals are favoured, the more interesting areas of the search space are analysed and the good features of the parents are extended to the whole population. Moreover, the mixing of genes allows the population to overcome local minima, improving the robustness of the algorithm.

Over successive generations, the population 'evolves' towards an optimal solution. The evaluation function, or objective function, provides a measure of performance with respect to a particular set of parameters. The fitness function transforms that measure of performance into an allocation of reproductive opportunities. The evaluation of a string representing a set of parameters is independent of the evaluation of any other string. The fitness of that string, however, is always defined with respect to other members of the current population [29].

As far as the basic structure of GAs is concerned, it should be kept in mind that the genes from a few comparatively highly fit (but not optimal) individuals may rapidly come to dominate the population, causing it to converge onto a local minimum or maximum. Once the population has converged, the ability of the GA to continue to search for better solutions is effectively eliminated: crossover of almost identical chromosomes produces little that is new. Only mutation remains to explore entirely new ground, and this simply performs a slow, random search [29]. According to [29], any efficient optimization algorithm must use two techniques to find a global maximum: *exploration* to investigate new and unknown areas in the search space, and *exploitation* to make use of knowledge found at points previously visited to help find better points. These two requirements are contradictory, and a good search algorithm must find a trade-off between the two.

Both GAs and ANNs are adaptive, learn, can deal with highly non-linear models and noisy data, and are robust, 'weak' random search methods. Neural networks and GAs demonstrate powerful problem-solving ability. They are based on quite simple principles, but take advantage of their mathematical nature: non-linear iteration [41]. GAs are often used in combination with ANN models (or directly with numerical models) in order to create a hybrid model with optimization abilities. Figure 4.10 shows the flow chart of a hybrid ANN–GA optimization model for the brake actuation pressure in a braking cycle, for example. This integrated approach to optimization can be summarized

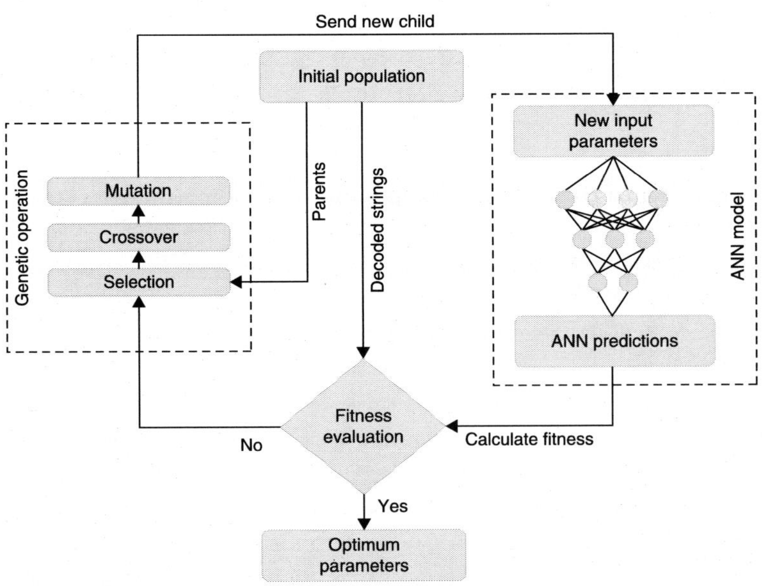

Figure 4.10 **A hybrid ANN–GA optimization model [12]**

in five main steps. (1) The GA generates an initial population of individuals for the applied brake pressure. (2) The fitness of each chromosome is evaluated. Also, a fitness value for each chromosome is assigned by the ANN model developed. (3) A population for the next generation is created by genetic operations (selection, crossover and mutation), with probabilities based on the fitness of each chromosome. (4) The GA then sends the new child generation to the ANN model as a new input parameter. (5) Finally, the fitness scores are calculated by the ANN model developed. The last four steps are repeated until a termination criterion is met. Obviously, the optimization process exploits the ability of the GA to optimize the neural-network inputs using the dynamic neural network model as its fitness function. As shown in Figure 4.9, the fitness of each individual is calculated by a dynamic neural model in each evolutionary step. For instance, by simulating the neural model for different speeds and temperatures, the brake actuation pressure in a braking cycle can be optimized. This approach takes into consideration dynamic changes in the real and wanted values of the braking torque and also non-linear constraints defined on the basis of these values. The definition of the constraints is very important. For instance, if the constraints are defined using the difference between the predicted and wanted braking torques, the wanted braking torque could be set differently. GAs can be used to search for the optimal value of the brake actuation pressure during a braking cycle in order to minimize the difference between the real and wanted values of the braking torque. The main goal is that the difference between the real and the wanted braking torque converges to zero in a series of successive generations of individuals. As the value of the fitness function, over successive generations, approaches the criteria that have been set, the applied pressure converges to the optimal value.

4.4 Conclusions

Artificial neural networks belong to a group of information-processing techniques which can be used to find knowledge, patterns or models from a large amount of data. Furthermore, intriguing advantages can be achieved by combining artificial neural networks with other computational models (FDM, FEM, FVM), which can provide the data to train the artificial neural network in order to create the model. Genetic algorithms can be used efficiently to find a suitable solution to a complex optimization problem. Genetic algorithms are an effective tool that is available for exploring large, complex search spaces based on Darwin's theory of survival of the fittest. In recent years, a lot of work has been presented combining genetic algorithms and artificial neural networks in the context of the development of hybrid methods. The main approaches that have been used in that sense are the following: (i) using genetic algorithms to improve the learning of artificial neural networks, and (ii) using genetic algorithms to perform a global search and to optimize the inputs of the neural network model versus the output results created by a network.

4.5 References

[1] Ćirović V., Aleksendrić D., Smiljanić D. Longitudinal wheel slip control using dynamic neural networks, *Mechatronics*, 23, 135–46 (2013).

[2] Sinha K., Krishnan R., Raghavendra D. Multi-objective robust optimization for crashworthiness during side impact, *International Journal of Vehicle Design*, 43, 116–35 (2007).

[3] Masten K.M. Electronics: The intelligence and intelligent control, *Annual Reviews*, 22, 1–11 (1998).

[4] Aleksendrić D., Duboka Č. Prediction of automotive friction material characteristics using artificial neural networks – cold performance, *Wear*, 261(3–4), 269–82 (2006).

[5] Abdi H. Linear algebra for neural networks, in Smelter N.J., Baltes P.B. (eds), *International Encyclopedia of the Social and Behavioral Sciences*, Elsevier Science, Oxford (2001).

[6] Aleksendrić D. Artificial neural networks in automotive brakes friction material development, Doctoral thesis, Faculty of Mechanical Engineering, University of Belgrade (2007).

[7] Miljković Z., Aleksendrić D. *Artificial Neural Networks – Solved Examples with Theoretical Background*, University of Belgrade Faculty of Mechanical Engineering (in Serbian) (2009).

[8] Simon J.R., Cole D. Modelling nonlinear vehicle dynamics with neural networks, *International Journal of Vehicle Design*, 53, 260–87 (2010).

[9] Aleksendrić D. Intelligent braking – technology, performance and economic challenge, in Zajac M., Nowaczek R. (eds), *Airports and the Automotive Industry: Security Issues, Economic Efficiency and Environmental Impact*, pp. 33–64, Nova Science Publishers, New York (2013).

[10] Ćirović V., Aleksendrić D., Mladenović D. Braking torque control using recurrent neural networks, *Proceedings of the Institution of Mechanical Engineers, Part D: Journal of Automobile Engineering*, 226, 754–66 (2012).

[11] Aleksendrić D., Duboka C. Fade performance prediction of automotive friction materials by means of artificial neural networks, *Wear*, 262, 778–90 (2007).

[12] Aleksendrić D., Senatore A. Optimization of manufacturing process effects on brake friction material wear, *Journal of Composite Materials*, 46(22), 2777–91 (2012).

[13] Cirovic V., Smiljanić D., Aleksendrić D. Neuro-genetic optimization of disc brake performance at elevated temperatures, *FME Transactions*, 42, 142–9 (2014).
[14] Xiao G., Zhu Z. Friction materials development by using DOE/RSM and artificial neural network, *Tribology International*, 43, 218–27 (2010).
[15] Aleksendrić D., Ćirović V. Neuro-genetic optimization of disc brake speed sensitivity, *International Journal of Vehicle Design*, 66 (3), 258–71 (2014).
[16] Aleksendrić D., Ćirović V. Smart brakes – neuro-genetic control of brake actuation pressure, in Bacciga A., Nalito R. (eds), *Recent Advances in Artificial Intelligence Research*, pp. 85–102, Nova Science Publishers, New York (2013).
[17] Mitchell A.M. *Computational Methods in Partial Differential Equations*, John Wiley and Sons (1969).
[18] Mitchell A.M., Griffiths D.F. *The Finite Difference Method in Partial Differential Equations*, John Wiley and Sons (1980).
[19] Smith G.D. *Numerical Solution of Partial Differential Equations: Finite Difference Methods*, Oxford University Press (1985).
[20] Leverque R.J. *Finite Difference Methods for Ordinary and Partial Differential Equations: Steady-state and Time-dependent Problems*, SIAM (2007).
[21] Hughes T.J.R. *The Finite Element Method: Linear Static and Dynamic Finite Element Analysis*, Prentice Hall, Englewood Cliffs, NJ (1987).
[22] Bathe K.J. *Finite Element Procedures*, Prentice Hall, Upper Saddle River, NJ (1996).
[23] Reddy, J.N. *An Introduction to the Finite Element Method*, McGraw-Hill (2005).
[24] Zienkiewicz O.C., Taylor R.L., Zhu J.Z. *The Finite Element Method: Its Basis and Fundamentals*, Butterworth-Heinemann (2013).
[25] Patankar S.V. *Numerical Heat Transfer and Fluid Flow*, Hemisphere (1980).
[26] LeVeque R.J. *Numerical Methods for Conservation Laws*, ETH Lectures in Mathematics Series, Birkhäuser (1990).
[27] Toro E.F. *Riemann Solvers and Numerical Methods for Fluid Dynamics*, Springer (1999).
[28] Eymard R., Gallouët T., Herbin, R. *Finite Volume Methods*, Handbook of Numerical Analysis, Vol. VII, pp. 713–1020, North-Holland, Amsterdam (2000).
[29] Busetti F. Genetic algorithms overview. Available from: *http://www.geocities.com/francorbusetti/gaweb.pdf* (2001).
[30] Whitley D. A. Genetic algorithm tutorial, *Statistics and Computing*, 4(2), 65–85 (1994).
[31] Kesgin U. Genetic algorithm and artificial neural network for engine optimization of efficiency and NOx emission, *Fuel*, 83, 885–95 (2004).
[32] Weise T. Global optimization algorithms – theory and application. Available from: *http://www.it-weise.de/* (2009).
[33] Wu M.H., Lin W., Duan S.Y. Developing a neural network and real genetic algorithm combined tool for an engine test bed, *Proceedings of the Institution of Mechanical Engineers, Part D: Journal of Automobile Engineering*, 220, 1737–53 (2006).

[34] Lee K.-M., Tsai J.-T., Liu T.-K., Chou J.-H. Improved genetic algorithm for mixed-discrete-continuous design optimization problems, *Engineering Optimization*, 42, 927–41 (2010).

[35] Thompson G.J., Atkinson C.M., Clark N.N., Long T.W., Hanzevack E. Technical Note: Neural network modelling of the emissions and performance of a heavy-duty diesel engine, *Proceedings of the Institution of Mechanical Engineers, Part D: Journal of Automobile Engineering*, 214, 111–26 (2000).

[36] Manan A., Vio G.A., Harmin M.Y., Cooper J.E. Optimization of aeroelastic composite structures using evolutionary algorithms, *Engineering Optimization*, 42, 171–84 (2010).

[37] Mashinchi M.R., Selamat A. An improvement on genetic-based learning method for fuzzy artificial neural networks, *Applied Soft Computing*, 9, 1208–16 (2009).

[38] Ko Y.D., Moon P., Kim C.E., Ham M.H., Myoung J.M., Yun I. Modelling and optimization of the growth rate for ZnO thin films using neural networks and genetic algorithms, *Expert Systems with Applications*, 36, 4061–6 (2009).

[39] Sahoo B., Maity D. Damage assessment of structures using hybrid neuro-genetic algorithm, *Applied Soft Computing*, 7, 89–104 (2007).

[40] Srinivasu D.S., Babu N.R. A neuro-genetic approach for selection of process parameters in abrasive water jet cutting considering variation in diameter of focusing nozzle, *Applied Soft Computing*, 8, 809–19 (2008).

[41] Koehn P. Combining genetic algorithms and neural networks: The encoding problem, Master of Science thesis, University of Tennessee, Knoxville (1994).

5

Composite materials – modelling, prediction and optimization

DOI: 10.1533/9781782421801.61

Abstract: Modelling, prediction and optimization of the performance of a composite material versus changes in its formulation and its manufacturing and operation conditions are important and challenging tasks in the framework of developing new composite materials. Owing to the complex synergistic effects of the formulation of a composite material and its manufacturing process on the actual performance of the material, an analytical model relating the basic features of a composite material to its in-service performance is quite difficult to obtain. That is why new computer-based techniques have to be employed in order to create an integrated environment for the modelling and optimization of the behaviour of composite materials under specific operation conditions. In this chapter, some soft computing techniques, such as artificial neural networks and genetic algorithms, as well as other computational methods (FEM, FDM, FVM), are used to model, predict and optimize the effects of the formulation and manufacture of a composite material on its final performance.

Key words: composite material, soft computing, modelling, prediction, optimization.

5.1 Introduction

Nowadays, composite materials play a crucial role in several engineering areas. The design and development of composite materials is a complex process owing to the fact that the effects of the formulation of a material and its manufacturing conditions are difficult to predict and to relate to its final in-service performance. An immediate practical example of this concept can be found by considering the composite materials that are aimed at application in motor vehicles. Today's motor vehicles have become considerably more complex owing to increasing requirements related to their active and passive safety, stability, manoeuvrability, etc. For instance, braking systems have always been given the highest importance concerning safety issues, with particular reference to active safety. Different design and load characteristics of passenger vehicles require different performance from their braking system. Furthermore, the performance of a braking system is strongly influenced by the complex interactions in the brake friction pair. The stochastic nature of the operation of friction brakes and their tribological behaviour are the main reasons why the operation of braking systems need to be constantly monitored and controlled depending on the driver's demands and the characteristics of the adhesion between the tyres and the road. The overall performance requirements imposed on a braking system, under a wide range of operating conditions, are high and manifold. They are particularly related to the brake friction pair. Regarding the interactions in the friction pair, the most important requirement relates to value of the coefficient of friction and its stability, particularly at elevated brake interface temperatures. The tribological processes occurring at the contact in the friction pair can span several different disciplines, such as mechanics, thermodynamics and chemistry. A dynamic change in the contact situation in a brake has a crucial influence on the friction behaviour of the friction pair and, accordingly, on the tribological performance of the brake. This means that the contact phenomena in the friction pair may be responsible for highly dynamic and stochastic variations in the braking torque during braking. To overcome this problem, the dynamic behaviour of brake operations need to be subjected to deep investigations in the form of modelling, prediction, optimization and control of the performance of disc brakes during a braking cycle. Providing possibilities for the prediction of the braking torque versus the brake actuation pressure, the sliding speed and the brake interface temperature is the main precondition for the future dynamic control of brake performance and for the development of 'smart brakes' [1]. Faster and better control of the braking process is extremely important in order to provide some kind of harmonization of the generated braking torque with the tyre–road adhesion conditions. The braking torque has a significant influence on the stopping distance, and this fact requires reliable performance of the brake friction material employed. So far, the development of new brake friction materials has mostly been done empirically because very little is known about the complex and highly non-linear phenomena that occur in the contact region during braking. Similar problems and difficulties could be said to exist for other composite materials used in motor vehicles.

This is the reason why analytical models of the behaviour of composite materials under different operation conditions are difficult or even impossible to obtain. In order to improve the performance of composite materials, a new approach to the modelling, prediction and optimization of the behaviour of composite materials is needed. The abilities of soft computing in the process of establishing complex non-linear functional relationships directly from experimental and numerical data have to be explored. Soft computing techniques offer the possibility that the process of modelling a composite material can be carried out without any prior assumptions about the nature of the input/output relationships. In this chapter, some possible applications of soft computing techniques will be shown. In particular, Section 5.2 describes some applications of numerical modelling and simulation-based computational optimization procedures in the manufacture of thermoset matrix composites. The capabilities of soft computing techniques in the context of the development of brake friction materials will be demonstrated in Section 5.3.

5.2 Numerical modelling and optimization of thermoset composite manufacturing processes

5.2.1 *Modelling issues*

Numerical modelling and simulation of the manufacturing processes of composites is a complex task, owing to the mutual interactions between the physical, chemical and mechanical phenomena involved and, of course, the intrinsic non-homogeneity and anisotropy of the materials. Even though some of the manufacturing processes for thermoset composites are conceptually simple and quite easy to understand, the analysis of the inherent dynamics and the definition of the optimal processing parameters are not trivial tasks, and require proper consideration of heat and mass transfer, species conversion, stress–strain development and porosity evolution [2]. The complexity of the problem can be seen by considering a generic liquid composite moulding process. At the beginning of the process, a fibre preform is placed in a die and possibly deformed to match the cavity geometry. As a consequence, local variations can be induced in some relevant properties of the preform, such as the permeability and conductivity, influencing the evolution of the resin flow front during impregnation and the heat transfer during the cure process. The resin flow is strongly affected by the resistance due to the porosity and permeability of the preform, as well as by the local viscosity of the resin being processed [3]. However, high-pressure viscous flows can induce further deformation of the fibre bundles [4]. Mass transport directly influences the distributions of the temperature and degree of cure, owing to conductive effects [5]. What is more, the evolution of the flow front and the pressure field, in conjunction with the temperature and degree of cure, conditions the diffusion, growth or dissolution of voids [3, 6–8].

Heat transfer and the resin reaction are strictly related to each other; indeed, the temperature experienced by the material being processed activates (and promotes to a certain extent) the resin reaction at different reaction rates (dependent on temperature and degree of cure). On the other hand, the heat generated by the exothermic cure reaction strongly influences the transient temperature distribution, the thermal peak experienced and the thermal gradients established [8]. Moreover, conductive heat flows are affected by the anisotropic nature of the composite material and therefore rely on the reinforcement architecture. Both aspects induce local viscosity variations, influencing the evolution of both saturated and unsaturated resin flow fronts [5]. Temperature and degree of cure are also extremely important in relation to process-induced stresses and strains, influencing first of all the state of the material (liquid, gelled solid or vitrified solid) and, as a consequence, its mechanical properties [9]. It is generally accepted that stresses and deformations in composite laminates are related to the interaction between the tool and the part [10], and to gradients in the temperature and degree of cure through the thickness of the composite [11]. Graphical evidence of the connections between the phenomena involved is depicted in Figure 5.1, well highlighting the multi-physics nature of the problem and models considered. Of course, the level of mutual interaction between the aforementioned aspects varies significantly from one manufacturing process to another, allowing one in some cases to adopt simplifying assumptions and reduce the theoretical complexity and the computational cost of the model considered. In the following paragraphs, the theoretical bases of the phenomena of interest are briefly recalled.

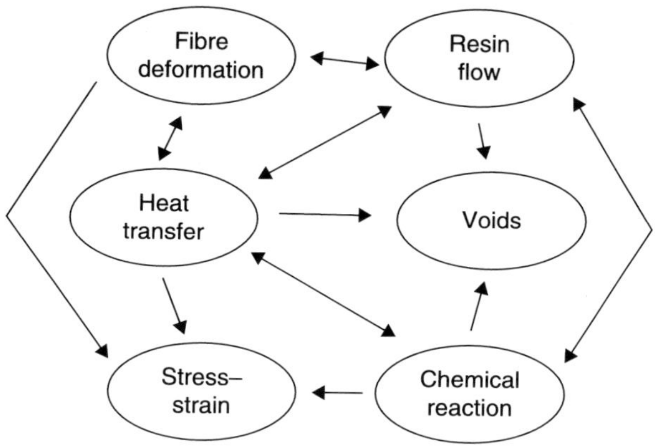

Figure 5.1 **Modelling issues and reciprocal interactions in composite-manufacturing processes**

Flow model

Resin flow is a common factor in most composite-manufacturing processes involving a thermoset resin. In autoclave processing, for instance, excess resin (in the liquid phase) should be removed from the impregnated preform and driven to a bleeder in order to maximize the fibre volume fraction and the mechanical properties of the final product. In conventional pultrusion systems, based on the open-bath technique for fibre wetting, the convergent inlet of the forming and curing die forces the resin to penetrate inside the fibre tows, reducing the composite porosity, and prevents the excess resin from entering the die, establishing a resin backflow. The relevance of all aspects of resin flow is unquestionable in all liquid composite processes, where it is necessary to completely wet the reinforcement, avoiding both macro- and micro-voids. The existence of a generalized theory to describe the flow of resin in composite processing was discussed in [12], concluding that each process can be modelled as a special case of Darcy's law, including specific assumptions (absence of fibre bundle deformations or isothermal and non-reactive flows, for instance) or specific effects, such as capillarity or relative permeability. Taking into account the fact that resin systems behave as non-compressible fluids, the governing equations for a generic single-phase or multiphase flow can be derived from the Navier–Stokes equations, in the form of continuity and momentum conservation laws. In particular, assuming that the pores in a fibre preform behind the moving flow front are fully saturated with resin, the flow of the liquid resin can be modelled using Darcy's law, as follows:

$$\nabla \cdot u = 0, \tag{5.1}$$

$$u = -\frac{K}{\mu} \cdot \nabla P, \tag{5.2}$$

where u is the velocity vector, μ represents the viscosity of the resin, K is the permeability tensor of the porous medium and P is the pressure. Combining Eqs 5-1 and 5.2 yields

$$\nabla \cdot \left(\frac{K}{\mu} \cdot \nabla P \right) = 0. \tag{5.3}$$

It should be noted that the set of equations described above is adequate only if the reinforcement can be represented as a porous medium with a uniform pore size distribution, or, in other words, as a single-scale porous medium. In most cases, the dual-scale nature of the fibre preforms typically used to realize advanced composite products cannot be neglected if one is to ensure reliability of the results obtained. Common reinforcing fabrics generally consist of several woven or stitched tows or bundles, which are made of thousands of individual fibres. The inter-fibre distance inside a tow is comparable to the fibre radius (a few micrometres), whereas the inter-tow spacing is of the order of millimetres. As a consequence, the resin flow is faster through the gaps (inter-tow channels) and the impregnation of the tows is delayed with respect to the position of the (unsaturated) flow front. The term 'dual-scale porous medium' implicitly suggests the coexistence of macro-scale (inter-tow) and micro-scale (intra-tow) flows and the need for two porosity values (inter-tow and intra-tow) and two permeability tensors (macro- and micro-permeability). What is more, the basic assumption of complete saturation of the porous medium behind the resin front applies only to the micro-flow; therefore some modifications of the mathematical model are needed.

In the interests of clarity, it may be relevant to make a distinction between two possible processing scenarios, representing preform impregnation (in a liquid composite moulding process, for instance) and the removal of excess resin (during autoclave processing). In the impregnation case, two approaches are generally adopted to deal with the dual-scale nature of the preform. The first is based on the solution of two sets of the above equations, particularized using values for the tow or the bulk, but without further modification. From the computational point of view, this strategy is applied by connecting 1D elements (representing the fibre tows) to the nodes of the mesh discretizing the whole geometry [13]. In the second approach, the validity of Eq. 5.3 is assumed only to describe the micro-flow, while on the macro-scale, Eq. 5.1 is modified by including a sink term (S) which accounts for the liquid resin that has penetrated into the fibre bundles [14]. One finally has

$$\nabla \cdot u = -S, \quad [5.4]$$

$$u = -\frac{K}{\mu} \cdot \nabla P, \quad [5.5]$$

$$\nabla \cdot \left(\frac{K}{\mu} \cdot \nabla P \right) = S. \quad [5.6]$$

A simpler situation is encountered in the latter case. Indeed, the difference in the velocity fields between the macro- and the micro-flow allows one to assume that the resin in the intra-tow region is stationary [14], reducing the problem to the single-scale case.

Heat transfer model

Heat transfer and the resin reaction are key issues in composite manufacturing, as briefly anticipated above. Generally speaking, the heat transfer problem in the processing of a composite material should be formulated using a non-thermal-equilibrium model, in order to account for the temperature difference (and, as a consequence, for the heat fluxes) between the fibres and the surrounding resin [5, 15, 16]. It should be noted that, in most cases, convective effects due to resin flow (as in liquid composite moulding) or fibre movement (as in pultrusion) cannot be excluded from the energy balance. As a consequence, besides the continuity and momentum equations for the fluid phase, one energy balance equation for each component is needed, including the heat transferred between contiguous phases. In a Cartesian coordinate system, neglecting the effects of voids and porosity, as well as variations in material properties, the temperature field can be obtained by solving the following equations for the liquid resin and the solid fibres, respectively:

$$\varphi \rho_r c_{p,r} \frac{\partial T_r}{\partial t} + \rho_f c_{p,r} u_r \cdot \nabla T_r = \varphi k_r \cdot \nabla^2 T_r + \varphi q + Q_{fr}, \quad [5.7]$$

$$\varphi_f \rho_f c_{p,f} \frac{\partial T_f}{\partial t} + \rho_f c_{p,f} u_f \cdot \nabla T_f = \varphi_f k_f \cdot \nabla^2 T_f + Q_{rf}, \quad [5.8]$$

where T is the temperature, t is the time, ρ is the density, c_p is the specific heat capacity, k is the thermal conductivity tensor and u is the velocity vector; the subscript 'r' refers to the resin and 'f' to the fibres. In the above equations, $\varphi = 1 - \varphi_f$ represents the volume porosity of the medium (the ratio between the volume available for fluid flow and the total volume). Assuming the absence of voids, φ coincides with the resin volume fraction $V_r = 1 - V_f$. Equations 5.7 and 5.8 are intimately coupled, since the interfacial heat transfer between the fluid and the solid is opposite to the interfacial heat transfer between the solid and the fluid, and depends on the temperature difference, the interfacial area density and the physical properties of the two phases:

$$Q_{rf} = -Q_{fr}. \quad [5.9]$$

A further source term q must be included on the right-hand side of Eq. 5.7 to account for the heat generated by the exothermic cure reaction of the resin [8]. This source term can be expressed as a function of the resin reaction rate R_r (the variation of the degree of cure with respect to time) and the total heat of reaction H_{tr}, obtainable by means of differential scanning calorimetry (DSC) measurements, as follows:

$$q = \rho_r R_r H_{tr}. \quad [5.10]$$

Significant simplifications of the thermal problem can be achieved under specific assumptions, whose validity relies on the manufacturing process modelled. Indeed, if the impregnation and the curing steps are not coupled, i.e. if the reaction is not activated during the preform wetting and resin flows are not influential during the solidification of the composite, the hypothesis of thermal equilibrium between resin and fibre becomes acceptable and the problem can be formulated considering only one energy equation (particularized using lumped material properties) without convective terms on the left-hand side, as follows:

$$\rho_c C_{p,c} \frac{\partial T_c}{\partial t} = k_c \cdot \nabla^2 T_c + \varphi q. \quad [5.11]$$

Chemical/rheological models

A fundamental step of any composite-manufacturing process is the consolidation or solidification of the workpiece, obtained by means of the exothermic cure or polymerization reaction of the matrix. The extent of the cure reaction is generally quantified by the so-called degree of cure. The main effect of this reaction is the establishment of several cross-links between the reactive monomers, accompanied by two relevant phase transitions, namely gelation and vitrification. These phenomena define the liquid–rubbery–solid transitions of the resin, as represented by the time–temperature–transformation (TTT) diagram depicted in Figure 5.2. What is more, the remarkable increase in the molecular weight of the resin structure induces a significant increase in the viscosity and other mechanical properties of the resin.

The degree of cure is a measurement factor of the conversion, rigorously defined as the ratio between the amount of reacted species and the total amount of reactive species. It can be evaluated experimentally following two different classes of methods, based on chemical or physical approaches. The analysis techniques belonging to the former class, such as Fourier transform infrared spectroscopy, nuclear magnetic resonance and high-performance liquid chromatography, are substantially direct methods, since the extent of the reaction is evaluated according the aforementioned rigorous definition. In contrast, physical methods, such as DSC, thermo-mechanical analysis and dynamical mechanical analysis, are indirect methods, since the extent of the reaction is computed by considering related phenomena. For instance, using DSC techniques, the instantaneous degree of cure is evaluated as the ratio between the amount of heat $H(t)$ released by the reactive material up to the time instant t considered and the total heat of reaction H_{tr}, as follows:

$$\alpha = \frac{H(t)}{H_{tr}}. \qquad [5.12]$$

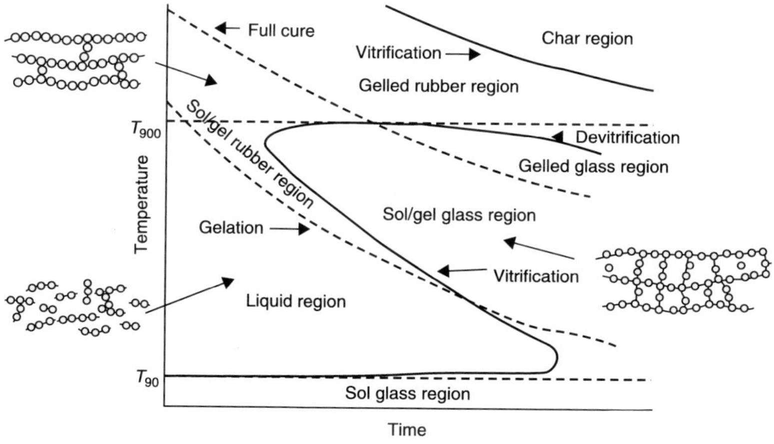

Figure 5.2 Time–temperature–transformation diagram for a generic thermoset resin

Several kinetic models can be found in the literature to describe the cure process [17–21]. Generally, kinetic models relate the rate of the resin reaction R_r (equal to the first derivative of the degree of cure with respect to time) to the temperature and degree of cure, according to the following equation:

$$R_r(\alpha, T) = K(T) \cdot f(\alpha), \quad [5.13]$$

where, for the first factor on the right-hand side of the equation, an Arrhenius dependence on the absolute temperature is commonly adopted:

$$K(T) = K_0 \exp\left(-\frac{E}{RT}\right) \quad [5.14]$$

Here, K_0 is a pre-exponential coefficient, E is the activation energy of the cure reaction and R is the gas constant.

Assuming that the overall conversion can be described by a unique relation, i.e. excluding the case of multiple exothermic peaks, the kinetic models commonly adopted are generally derived from the following two forms:

$$R_r(\alpha, T) = f(\alpha) \cdot (K_1 + K_2\, \alpha^m)\, (1 - \alpha)^n \quad [5.15]$$

or

$$R_r(\alpha, T) = f(\alpha) \cdot K \cdot \alpha^m\, (1 - \alpha)^n, \quad [5.16]$$

which are valid, respectively, for isothermal or dynamic curing processes. The term $f(\alpha)$ represents a diffusion factor, which accounts for a transition (from kinetic to diffusive) in the reaction behaviour. It is usually written as

$$f(\alpha) = \frac{1}{1 + \mathrm{e}^{C(\alpha - \alpha_c)}}, \quad [5.17]$$

where C and α_c are temperature-dependent parameters.

During the manufacture of a composite, such as in liquid composite moulding processes and pultrusion, the resin flows through a porous preform or moves with respect to tools. As a consequence, the degree of cure can be considered as a generic transport variable, which, neglecting diffusive effects, obeys the following transport equation:

$$\frac{\partial \alpha}{\partial t} + u \cdot \nabla \alpha = R_r. \quad [5.18]$$

Rheological models aim to describe the variation of the resin viscosity μ as a function of the temperature and degree of cure, as well as the onset of gelation and vitrification, the gel time, and the glass transition temperature T_g. Resin viscosity is a very important parameter during the impregnation or compaction of a preform. Generally speaking, early activation of the cure reaction, which induces a sharp viscosity increase and gelation of the resin, should be avoided, in order to completely wet the fibres, to create a continuous interphase and to let entrapped volatiles escape. In the case of dynamic curing processes, the local viscosity can be predicted according to, for instance, the following rheological model [22]:

$$\mu = A_\mu \cdot e^{\left(\frac{E_\alpha}{RT}\right)} \cdot \left(\frac{\alpha_g}{\alpha_g - \alpha}\right)^{A+B\alpha} \qquad [5.19]$$

or, alternatively, according to the well-recognized three-parameter correlation model [23], which is expressed as

$$\mu = \mu_\infty \exp\left(\frac{E_\mu}{RT} + K\alpha\right), \qquad [5.20]$$

where α_g is the degree of cure at gelation (the gel point), R is the gas constant, T is the absolute temperature, and μ_∞, E_μ, K, A and B are material parameters provided by experimental data fitting. It should be noted that after the gel point, the dramatic viscosity increase inhibits further resin flow and the above equations are not effective any more. From a practical point of view, when one is planning a generic manufacturing process involving reactive thermoset resins, it is relevant to consider the time available for resin flow, or, in other words, the interval of time from the mixing of the resin with the catalyser to the flow stop condition. This time is defined taking into account the gelation time t_{gel}, i.e. the time needed for the cure reaction to reach a degree of cure equal to α_g Assuming the general definition of the reaction rate provided in Eq. 5.14, this results in

$$t_{gel} = \frac{1}{k(T)} \int_0^{\alpha_g} \frac{1}{f(\alpha)} d\alpha. \qquad [5.21]$$

Unlike gelation, the glass transition is a reversible phenomenon, manifested as the resin temperature passes (on cooling) through an unequivocally degree-of-cure-dependent temperature [24, 25], namely the glass transition temperature T_g, represented on the TTT diagram by an S-shaped curve. Under isothermal conditions, the resin is in a glassy state if the degree of cure is higher than the value matching the glass transition condition at that temperature. The glass transition is accompanied by a significant increase in the stiffness of the resin (and the composite), which influences, as a consequence, the evolution of the stresses and strains. The inverse transformation is called devitrification. The most commonly adopted analytical equation relating T_g and α is the modification proposed by Pascault and Williams [26] of the Di Benedetto equation [27]

$$T_g = \frac{(1-\alpha)T_{g0} + \lambda\alpha\, T_{g\infty}}{(1-\alpha)+\lambda\alpha}, \qquad [5.22]$$

where T_{g0} and $T_{g\infty}$ are the glass transition temperatures for the uncured ($\alpha=0$) and totally cured ($\alpha=1$) resin systems and λ is a coefficient in the interval [0, 1] depending on the structure of the resin. What is more, the glass transition alters the behaviour of the cure reaction, whose behaviour switches from kinetic to diffusive, with a drastic decrease in the reaction rate.

Stress–strain model

Process-induced stresses and strains must be properly considered in order to effectively assess the in-service loading scenarios and performance of a structural composite product. It is widely accepted that the main factors affecting the transient and residual stress fields are the tool–workpiece interaction, due to mechanical contact and friction phenomena, and thermal expansion and chemical shrinkage. As usual, the formulation of the mechanical properties can be based on an energy balance, which can be represented by the principle of virtual work. A crucial point in this kind of model is the appropriate definition of the mechanical behaviour, i.e. the constitutive model, of the thermoset composite material during the manufacturing process. The phase changes (from liquid to gelled solid and, finally, to the glassy state) of the resin system during curing, in conjunction with the interactions between the matrix and the reinforcement, make this question quite complex indeed. In this context, a useful simplification can be achieved by adopting a quasi-static approach and assuming a linear relationship between stresses and strains, where the stiffness matrix is adjusted step by step, taking account of the variation of the modulus of the resin with the temperature and the degree of cure. Some models have been presented and discussed in the literature to account for this dependence [28–31]. Two different models have been derived in [32]. The first is basically an adaptation, accounting for the effects of temperature, of the model discussed in the pioneering reference work by Bogetti and Gillespie [29]. The modulus of the resin is assumed to vary between the modulus E_r^0 of the uncured resin (degree of cure α equal to zero) and the modulus E_r^∞ of the fully cured resin (degree of cure equal to one), according to the following equation:

$$E_r = E_r'[1 + \alpha_{Er}(T - T_0)], \qquad [5.23]$$

where T and T_0 are the local temperature and the reference temperature for modulus development of the resin system, respectively, and α_{Er} is a coefficient expressing a secondary dependence on T. The degree-of-cure dependence is included in the equations providing the parameter E_r', as follows:

$$E_r' = \begin{cases} E_r^0 & \alpha < \alpha_{C1} \\ (1-\alpha_{mod})\, E_r^0 + \alpha_{mod} E_r^\infty + \gamma_r \alpha_{mod}(1-\alpha_{mod})(E_r^\infty - E_r^0) & \text{for } \alpha_{C1} \le \alpha \le \alpha_{C2} \\ E_r^\infty & \alpha > \alpha_{C2} \end{cases}. \qquad [5.24]$$

In the above equations, α_{C1} and α_{C2} represent the degrees of cure at which the resin modulus begins to develop and reaches its final value, respectively; γ_r is a factor for the hardening rate of the resin; and α_{mod} is defined as

$$\alpha_{mod} = \frac{\alpha - \alpha_{C1}}{\alpha_{C2} - \alpha_{C1}}. \qquad [5.25]$$

The second approach, named the Cure Hardening Instantaneous Linear Elastic (CHILE) approach, was proposed by Johnston [32]. It assumes the same relation between the parameter E_r' and the temperature T as in Eq. 5.23; however, E_r' is expressed as a function of a fictitious temperature T^*, as follows:

$$E_r' = \begin{cases} E_r^0 & T^* < T_{C1} \\ E_r^0 + \dfrac{T^* - T_{C1}}{T_{C2} - T_{C1}}(E_r^\infty - E_r^0) & \text{for } T_{C1} \le T^* \le T_{C2} \\ E_r^\infty & T^* > T_{C2} \end{cases}. \qquad [5.26]$$

The fictitious temperature T^* is defined as the difference between the degree-of-cure-dependent glass transition temperature T_g and the actual resin temperature T. T_{C1} and T_{C2} are the critical temperatures, defining the beginning and the end of modulus development. Once the elastic properties of the resin have been computed, a further step is required to evaluate the lumped or homogenized mechanical properties of the composite undergoing processing, which are, of course, dependent on the specific properties, architecture and volume fraction of the reinforcement.

Multi-scale analysis and stochastic effects

Advanced composites, being a combination of a continuous phase (the matrix) and a discontinuous strongly orientated phase (the reinforcement), are intrinsically non-homogeneous and anisotropic materials. In several cases, fibre architectures characterized by great complexity of the arrangement of tows compared with unidirectional laminates are employed, in order to balance the mechanical properties along different (generally in-plane) directions, preserving the advantageous specific properties of the composite. Typical examples of these reinforcement structures are woven, braided and crimped fabrics. The effective application of these configurations, however, is subordinate to the accurate characterization not only of the mechanical properties (at the design stage) but also of physical and other properties, such as the conductivity and the permeability, which are of great interest when planning the manufacturing process parameters. Experimental testing has a dominant role in both academic and industrial practice; however, despite the relevance of (and unavoidable need for) experimental results, these measurements are case-specific and do not have predictive capabilities.

As a consequence, in recent years, remarkable research effort has been spent on the development of numerical models to evaluate the properties of composite materials, taking into account the properties of each component. In this framework, a great deal of attention has recently been focused on multi-scale approaches [33–45], based on a coupling between analyses and simulations performed by considering the same material or element on different length scales, as depicted in Figure 5.3. The basic concept of multi-scale modelling is to transfer the outcomes obtained by analysing a unit cell or representative volume element (RVE) at a lower scale to a higher scale, using appropriate homogenization techniques. In the case of composite materials, the micro-, meso- and macro-scales are defined by considering the fibre radius (micrometres), the tow dimensions (millimetres) and the product size (metres) as unit lengths (Figure 5.3).

The interest in multi-scale modelling is basically due to two relevant reasons: the possibilities of significantly reducing both the experimental effort and the computational cost with respect to single-scale simulations,

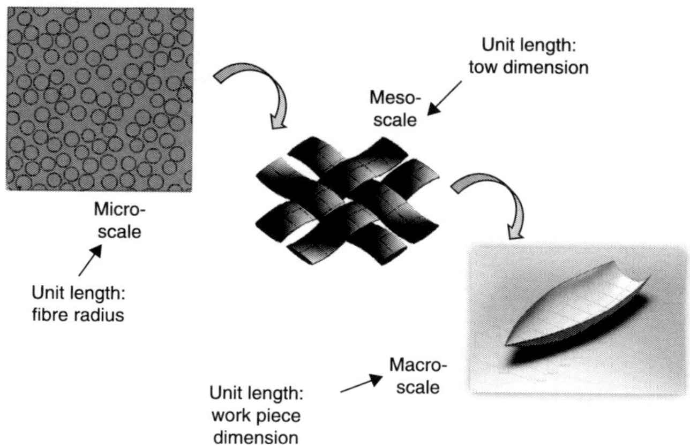

Figure 5.3 Micro-, meso- and macro-scales in composite-manufacturing simulation

since the modelling of each fibril, even at the meso-scale, is computationally unsustainable. What is more, multi-scale approaches also show the intriguing feature that is it relatively easy to include stochastic perturbations in the analysis.

Indeed, the influence of the deviation of the arrangement of fibres, with respect to a square or hexagonal packing, on the homogenized properties of an RVE at the micro-scale has been the focus of previous studies, aiming to explain the scatter in the experimental data. The opportunity for proper consideration of the geometric disorder of the fibre distribution in RVEs has been remarked on in [38, 39] and [40–43] in relation to calculations of mechanical properties and permeability, respectively.

The generation of the unit cell must satisfy two important prerequisites: it should be as small as possible in order to reduce the computational cost, but at the same time it should reproduce the actual behaviour of the material considered at the upper scale. Several algorithms have been proposed in the literature to generate RVEs respecting the aforementioned conditions for a broad range of fibre volume fraction, based on two different approaches. The first class of methods generates the perturbed RVE by randomly defining the position of each fibre's centre in the unit cell, respecting some constraints such as no overlap between fibres, the volume fractions of the components, or a minimum inter-fibre distance. Differently, algorithms of the second class start from an initial regular fibre arrangement in the RVE and then perturb the reference distribution by means of random displacements of each fibre, conditioning the acceptance of each movement using similar criteria to those above. A more accurate description, including an analysis of some statistical functions used to quantitatively characterize the spatial distribution of the fibres, can be found in [44, 45].

Regardless of the algorithm adopted for RVE generation, its statistical validity, relative to the size and fibre distribution, can be investigated using some statistical functions, such as the Morishita number I_δ, the Ripley K function and the pair distribution function g_2, briefly recalled in the following.

The Morishita index is a statistical measure of the dispersion, computed iteratively by dividing the cell considered into a progressively increasing number q of equally sized subsections having the same shape as the whole cell, and then counting the number of fibres n_i in each subsection [46], as schematized in Figure 5.4. It is analytically formulated as

$$I_\delta(q) = q\delta(q) = q\sum_{i=1}^{q} \frac{m_i(m_i - 1)}{N_f(N_f - 1)}. \qquad [5.27]$$

The Morishita number is obtained by assuming in the above equations the condition $q = N_f$, corresponding to equality between the number of subsections and the number N_f of fibres in the cell. In particular, a

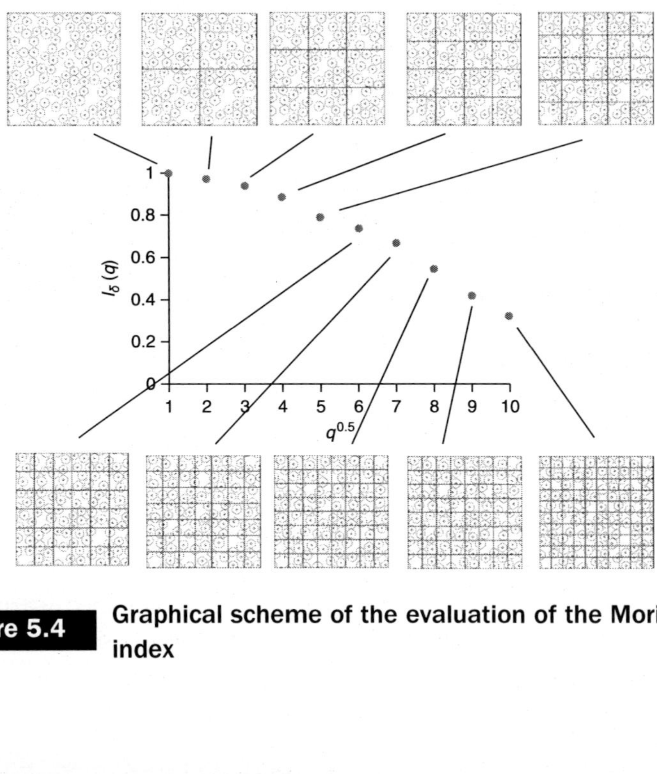

Figure 5.4 Graphical scheme of the evaluation of the Morishita index

Morishita number less than one indicates a regular, homogeneous fibre packing, a value of unity indicates a homogeneous random distribution, and larger values indicate agglomerate distributions.

In a similar fashion, the Ripley K function is generally used to discriminate regular, random and grouped distributions of points [47]. Assuming that the fibre distribution is defined by the centres of the fibres, the K function can be written in the following form:

$$K(r_k) = \frac{A}{N_f^2} \sum_{k=1}^{N_f} \frac{I_k(r_k)}{w_k(r_k)}, \qquad [5.28]$$

where A is the total area of the unit cell, $I_k(r_k)$ is the number of centres lying within a distance r_k from an arbitrary centre k, and $w_k(r_k)$ is a weight factor taking account of the fact that part of the observation area could fall outside the boundary of the unit cell. A graphical representation of the parameters defining the Ripley function is provided in Figure 5.5.

For data analysis, a linearized form of the Ripley K function (the L function) is commonly used, which can be written as

$$L(r) = \sqrt{(K(r)/\pi)}. \qquad [5.29]$$

This is generally used and compared with the same function related to the Poisson distribution, as in Figure 5.6. The L function of a stochastic distribution overlaps substantially with the L function of the Poisson distribution; higher values indicate agglomerate distributions, while lower values indicate a regular one.

Finally, another widely adopted descriptor is the pair distribution function g_2, which is defined as the probability of finding the centre of a fibre inside an annulus of internal radius r and thickness dr with its centre

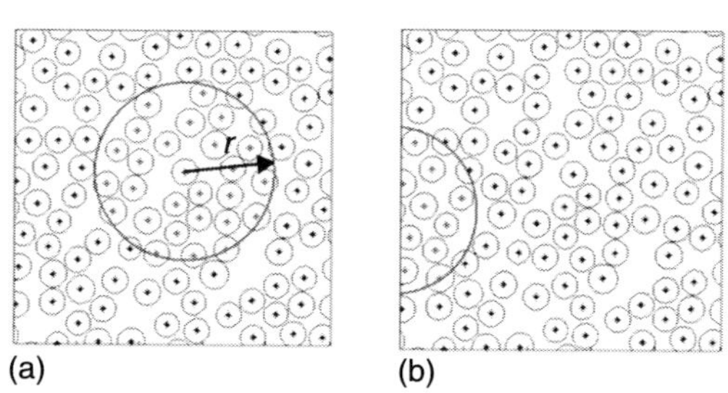

Figure 5.5 Graphical scheme of the evaluation of the Ripley function, showing the definition of $I_k(r)$ and the weight factor: (a) $w_k = 1$, (b) $w_k \neq 1$

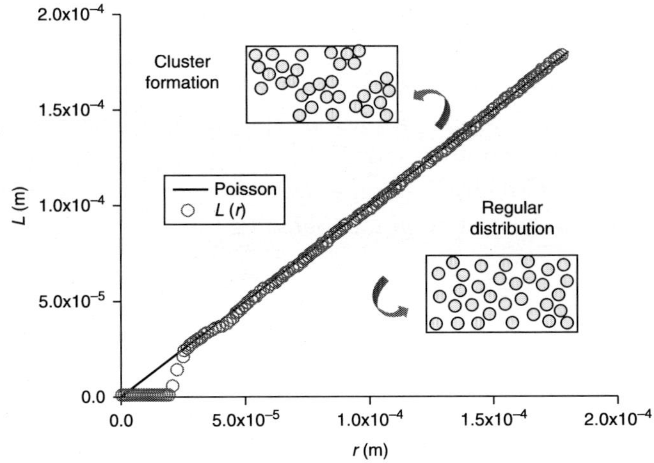

Figure 5.6 ***L* function: comparison with the Poisson distribution**

at a randomly selected fibre [48]. The pair distribution function describes the intensity of distances between fibres and can be written as

$$g_2(r) = \frac{1}{2\pi r_g} \frac{dK(r)}{dr}. \qquad [5.30]$$

This is equal to unity for a Poisson distribution, and its deviation from unity indicates the degree of spatial correlation between two points.

In the following paragraphs, some numerical applications of the above models and methods to selected manufacturing processes for thermoset matrix composites are shown and discussed.

5.2.2 Autoclave curing

Thermochemical–rheological model

The autoclave curing process is widely employed to manufacture high-temperature resin systems and to improve the mechanical properties and the maximum service temperature of composite structural products by increasing their degree of cure. The relevance of the curing conditions to the mechanical properties of the processed composite has been well highlighted in the literature [8, 49, 50]. Indeed, the activation of the cure reaction, the suppression of voids and the stress–strain evolution depend mainly on the temperature history.

A full model of the autoclave curing process involves several physics to be considered, in order to predict phenomena related to heat and mass transfer, monomer cross-linking, void growth and suppression, and the stress–strain distribution. As a consequence, the model should be implemented as a computational fluid dynamics (CFD) multiphase thermochemical model, including mesh deformations to account for the applied external pressure, coupled with a non-linear stress–strain analysis. The computational effort needed to obtain a solution to this problem is considerable and makes such a model unsuitable for implementation in simulation-based optimization procedures. However, a remarkable reduction in the complexity of the model and, as a consequence, in the computational time required can be achieved by assuming initially that the resin flow does not influence the evolution of the heat transfer or of the chemical reaction [28, 51–53]. The design of the cure cycle, with specific reference to the definition of the temperature–time curve, is a crucial issue for competitive production, and computationally assisted process planning is increasingly preferred to expensive trial and error procedures.

One simplified model is based on iterative solution of the energy equation with the species equation. In particular, assuming thermal equilibrium between the resin and the fibres, lumped material properties can be used. The energy balance equation, according to the first law of thermodynamics and the Fourier model, can be written as follows for a generic domain defined by a control volume V and a boundary surface S:

$$\rho_c c_{p,c} \frac{\partial T}{\partial t} = \nabla \cdot (K_c \nabla T) + u_{gen}, \qquad [5.31]$$

where ρ_c and $c_{p,c}$ denote the density and specific heat capacity of the composite material, K_c represents the thermal conductivity tensor, and u_{gen} is the rate of internal energy generation. Note that the material properties in Eq. 5.31 are assumed to be constant. Multiplying both sides of Eq. 5.31 by a virtual temperature increment δT and integrating over the domain considered yields

$$\int_V \rho_c c_{p,c} \frac{\partial T}{\partial t} \delta T \, dV = \int_V \nabla \cdot (K_c \nabla T) \delta T \, dV + \int_V u_{gen} \, \delta T \, dV. \qquad [5.32]$$

Taking into account that

$$\nabla \cdot (\delta T \; K_c \nabla T) = \nabla \delta T \cdot (K_c \nabla T) + \delta T \nabla \cdot (K_c \nabla T), \qquad [5.33]$$

it follows that

$$\int_V \rho_c c_{p,c} \frac{\partial T}{\partial t} \delta T \, dV = \int_V \nabla \cdot (\delta T K_c \nabla T) \, dV - \int_V \nabla \delta T \cdot (K_c \nabla T) \, dV$$

$$+ \int_V u_{gen} \delta T \, dV. \quad [5.34]$$

Indicating by q_n the thermal flow acting on a surface S characterized by the unit outward normal vector n, and taking into account the divergence theorem, Eq. 5.34 can be rearranged as

$$\int_V \rho_c c_{p,c} \frac{\partial T}{\partial t} \delta T \, dV + \int_V \nabla \delta T \cdot (K_c \nabla T) \, dV$$

$$= \int_s \delta T \, q_n \, dS + \int_V u_{gen} \, \delta T \, dV, \quad [5.35]$$

where the first term indicates the variation of the internal energy over time, the second is the conductive flux, the third is the external (convective or imposed) fluxes and the fourth is the heat generation rate due to the resin reaction. In several cases [28, 51–56], equivalent heat transfer coefficients accounting for all thermal resistances have been employed to define the external heat fluxes, in order to reproduce the effect of the tooling and the vacuum bag, as depicted in Figure 5.7.

Equation 5.35 represents a virtual-temperature principle to be solved numerically, assuming appropriate initial and boundary conditions. In the case described here, a finite element scheme was adopted.

Under the assumption of negligible resin flow, the conversion equation (Eq. 5.18) reduces to the following:

$$R_r = R_r(\alpha, T) = \frac{\partial \alpha}{\partial t}, \quad [5.36]$$

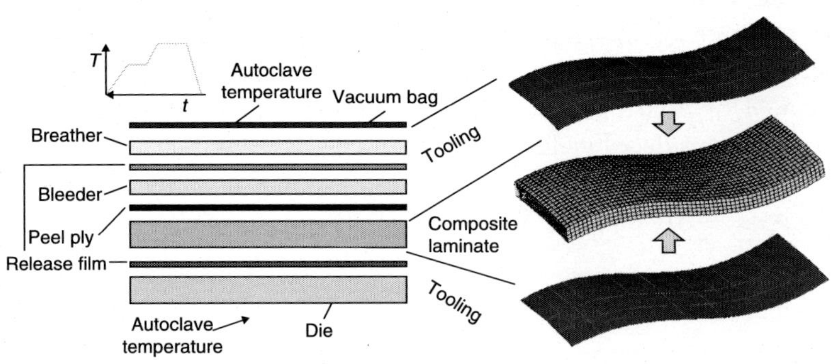

Figure 5.7 **Lay-up of a process and corresponding finite element three-dimensional scheme**

and, according to the definition of α as the ratio of the amount of heat $H(t)$ evolved during the curing process up to time t to the total heat of reaction H_{tr}, the generative term can be written as

$$u_{gen} = \rho_r V_r H_{tr} R_r, \tag{5.37}$$

where ρ_r is the resin density and V_r is the resin volume fraction.

For transient analysis, an effective procedure can be obtained by decoupling the heat transfer and the chemical problem with respect to time, by computing the degree of cure at each node i and at each time step m. The degree of cure and the reaction rate in the same node at the previous time step are considered, using a discretization of Eq. 5.36, which can be written, after some manipulations, as

$$\alpha_i^m = \alpha_i^{m-1} + R_r(\alpha_i^{m-1}, T_i^{m-1})\ \Delta t. \tag{5.38}$$

The degree of cure provided by Eq. 5.38 can be then used to update the reaction rate and the heat generation rate per unit mass in each node, which is taken as an internal load for the load step m; the above calculations are repeated until the end of the process.

The system of equations described above was used to simulate the curing process of a 50 mm thick graphite–epoxy laminate. The numerical results obtained were compared with corresponding outcomes provided by a coupled flow–thermochemical model [54, 55], as well as with experimental data, to analyse the validity of the simplifying hypothesis adopted. Taking into account that the in-plane dimensions of composite parts are generally much greater than the thickness, a one-dimensional model, along the thickness (z axis) direction, was adopted, as schematized in Figure 5.8. The finite element model of the laminate undergoing processing was built using 10 elements and 11 nodes.

The architecture of the fibre perform was assumed to be unidirectional along the larger of the two planar directions, modelled as the x axis. Carbon fibres and Hercules AS4/3501–6 epoxy resin were assumed as the reinforcement material and the polymeric matrix. The physical properties and concentrations of each constituent [54, 55] are reported in Table 5.1. The physical properties of a composite material undergoing processing can be evaluated using the well-known rule of mixtures, taking into account the reinforcement fibre architecture, the properties of the fibres and of the matrix, and the respective volume fractions.

Different equations were implemented to model the conversion rate, taking into account the multiple exothermic peaks shown by the resin system considered [54, 55, 57], in particular,

$$R_r = \begin{cases} (K_1 + K_2\alpha)(1-\alpha)(0.47-\alpha) & \text{for} \quad \alpha \le 0.3 \\ K_3(1-\alpha) & \qquad\quad \alpha \le 0.3 \end{cases}, \qquad [5.39]$$

where K_1, K_2 and K_3 are temperature-dependent parameters, following the Arrhenius equation

$$K_i = A_i \exp\left(-\frac{\nabla E_i}{RT}\right), \qquad i = 1, 2, 3. \qquad [5.40]$$

Here A_i is a pre-exponential constant, ΔE_i is the activation energy, T is the absolute temperature and R is the universal gas constant. In the model in

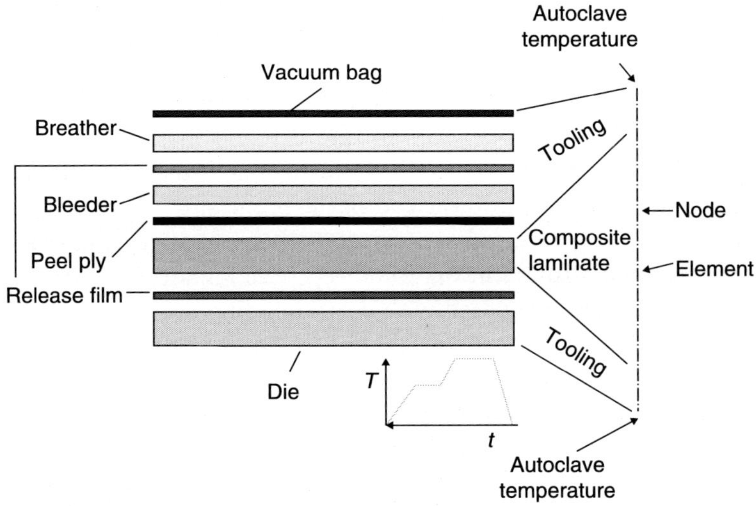

Figure 5.8 Lay-up of the process and corresponding finite element one-dimensional scheme

Table 5.1 Physical properties and concentrations of materials [54, 55]

Property	Graphite	Epoxy
ρ (kg m^{-3})	1790	1260
cp (J kg^{-1} K^{-1})	712	1260
kx (W m^{-1} K^{-1})	26.0	0.167
ky (W m^{-1} K^{-1})	2.60	0.167
kz (W m^{-1} K^{-1})	2.60	0.167
Volume fraction	0.62	0.38

Table 5.2 Parameters of resin kinetics

Parameter	Value
A_1 (min^{-1})	2.101×10^9
A_2 (min^{-1})	-2.014×10^9
A_3 (min^{-1})	1.960×10^5
ΔE_1 (J mol^{-1})	8.07×10^4
ΔE_2 (J mol^{-1})	7.78×10^4
ΔE_3 (J mol^{-1})	5.66×10^4
H_{tr} (J kg^{-1})	474×10^3

Eq. 5.39, an autocatalytic kinetic model describes the early stages of the cure process, where the reaction of epoxy groups with primary amine and hydroxy groups is prevalent; then a first-order model characterizes the successive stage of the cure process, where reactions between secondary amines and epoxy groups to form tertiary amines are manifested. The parameters [51, 52, 54, 55, 57] of the kinetics model adopted are reported in Table 5.2.

The possible onset of resin vitrification, with consequent modification of the reaction from kinetic to diffusive, was considered by imposing a null reaction rate in the nodes where a glass transition was detected. The glass transition temperature as a function of the degree of cure was defined following the relation proposed by Kim et al. [58] for the material considered. In each iteration, the reaction rate vanished (and the reaction stopped) if the computed nodal temperature was lower than the local glass transition temperature.

Finally, the solution of the system of equations described above allowed us to evaluate also the evolution of the resin viscosity during the process, delineating, to a first approximation, the process window from the beginning of the process until gelation of the resin. The parameters reported in Table 5.3 were used to characterize the three-parameter rheological model presented in Eq. 5.20.

Simulated results, i.e. profiles of the temperature, degree of cure and viscosity at the centre and on the surface of the composite laminate undergoing processing, are shown in Figures 5.9, 5.10 and 5.11 and compared with reference experimental and numerical data [57].

As can be seen in Figure 5.9, the temperature at the surface of the composite is higher than that at the core until the activation of the resin reaction (end of the first heating ramp), as visible also in Figure 5.10. Then,

Table 5.3 Parameters of the rheological model

Parameter	Value
μ_∞ (Pa s)	7.93×10^{-14}
$E\mu$ (J mol^{-1})	9.08×10^{4}
K	14.1
R (J K^{-1} mol^{-1})	8.314

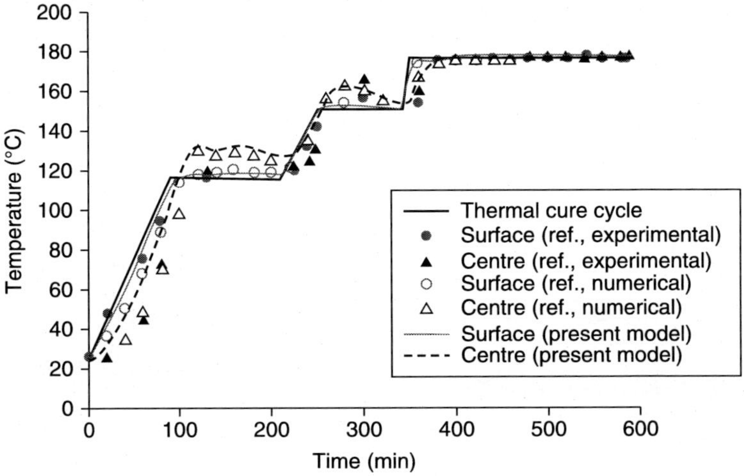

Figure 5.9 Temperature profiles: numerical results and reference data [57]

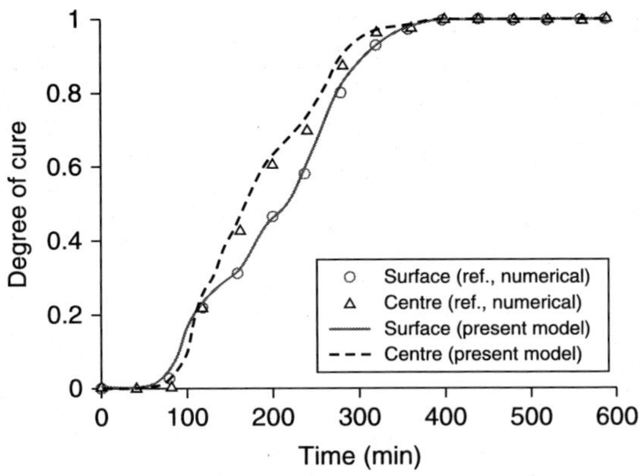

Figure 5.10 Degree-of-cure profiles: numerical results and reference data [57]

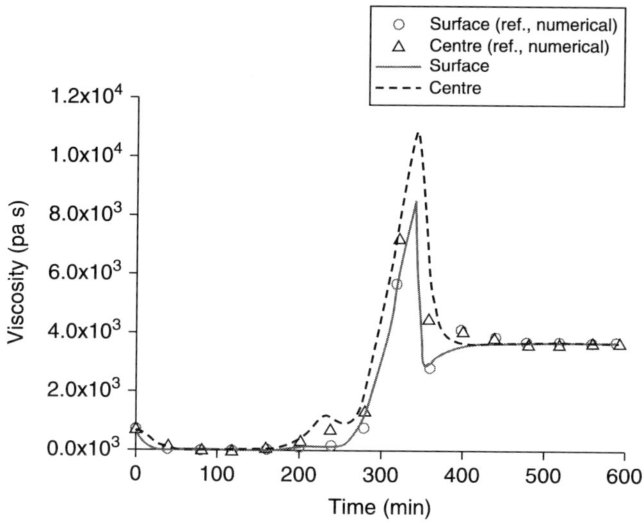

Figure 5.11 Viscosity profiles: numerical results and reference data [57]

strong heat generation results in the first exothermic peak (at about 130 °C) at the centre. The main effect of the first dwell at 115 °C is a reduction in the thermal gradient inside the material; however, the second heating ramp, up to 150 °C, promotes further cross-linking and heat dissipation, inducing a second thermal peak (at about 160 °C). The peak temperatures are higher in the core of the material, because the low conductivity of the resin system prevents the heat generated from diffusing towards the external surface. The third dwell, at 180 °C, allows the resin to reach full cure, avoiding the glass transition during the process. It should be noted that, since most of the resin has already reacted (Figure 5.10), the released heat is not enough to induce other peaks and the temperature on the heated surface prevails over the inner temperature. Figure 5.10 also shows that the cure reaction is activated earlier at the surface, owing to the external heat flux. However, after 120 minutes of processing, a cure crossover is detected at approximately $\alpha = 0.22$, evidencing the desired inside–out solidification [28].

The results reported here highlight the good superimposition of the temperature and degree-of-cure profiles on the reference data that was found in all of the simulated cycles, showing the validity of the assumptions used. In particular, excellent agreement between the data provided by the full and the present model is detected after the activation of the resin reaction (at about 100 min). Indeed, the increase in the degree of cure due to reaction activation leads to a fast increase in the resin viscosity (Figures 5.11 and

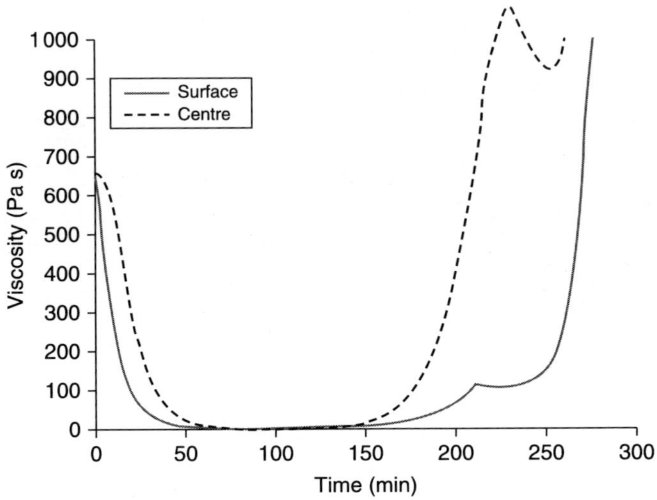

Figure 5.12 Viscosity profiles, showing the process window

5.12), which strongly reduces the resin flow and finally causes it to vanish, resulting in substantial equivalence of the full and simplified models.

The initial heating of the material leads to a reduction in the resin viscosity, allowing excess resin to flow under the action of the vacuum and the external pressure from the reinforcing layers towards the bleeder (Figure 5.12). As a consequence, the resin volume fraction, the physical properties of the material and the thickness of the product change dynamically until the onset of gelation and the flow stop condition. Relatively large differences between the temperature profiles can be observed at the beginning of the process; however, the influence of the error attributable to the hypothesis of no resin flow on the global solution can be considered as acceptable in practice, taking into account also the agreement of the degree-of-cure profiles reported in Figure 5.11.

5.2.3 Multi-scale stochastic analysis of the process

The intriguing capabilities of computational simulation models, as a complementary tool to experimental analysis to achieve time and cost compression in the planning of autoclave processes compared with classical trial and error procedures, are quite clear when one looks at the application described in Section 5.2.1. Similar models of the curing process, based mainly on the finite element method, have been proposed, validated and discussed in the literature [51–53, 56]. Most of the proposed approaches, however, rely on deterministic hypotheses, assuming perfect fibre packing and neglecting dimensional variations in the fibre radii, in strong contrast to experimental observations. The opportunity for proper consideration of geometric disorder in the fibre distribution has been remarked on in [38, 59–61], with respect to mechanical, physical and flow properties. In this framework, promising results can be achieved by means of a multi-scale (micro–meso–macro) approach to process modelling. In this section, some implications of the variations in thermal conductivity due to stochastic perturbation of the reinforcement are numerically investigated using a multi-scale non-deterministic model. Thermal simulations of the heat flow through perturbed RVEs were performed, by solving the energy balance and finally deriving the transverse conductivity by means of a Fourier model. The effect of stochastic perturbations was investigated by simulating the curing process of a 50 mm thick unidirectional composite laminate, taken as a benchmark. This fibre architecture implies coincidence of the material behaviour on the micro- and meso-scales, allowing one to take a shortcut between the micro-scale and macro-scale analyses. It should be noticed, however, that when one is considering a more complex reinforcement structure (as generally employed), a further analysis step of this kind is needed.

Micro-scale analysis

This step involves the generation of randomly perturbed RVEs, the simulation of the phenomenon of interest (heat transfer, as far as the thermal conductivity is concerned) and the homogenization of the results. The unit cells were obtained by inducing perturbations in an initial regular distribution by means of a generation code, graphically schematized in Figure 5.13, which was implemented as a script in the MATLAB environment.

The initial cell geometry was defined as a perfect (square or hexagonal) fibre packing, imposing the porosity φ or, equivalently, the fibre volume fraction V_f, the number of fibres N_f and the mean fibre radius r. This configuration was altered by introducing a random variation in the radius of each fibre, following a normal distribution centred on the mean value r, with a predefined standard deviation σ. An initial check was performed by considering a criterion based on the minimum distance d_{min} between adjacent fibres, defined as a fraction of the mean radius. More specifically, if the generation of an initial cell failed owing to violation of the above constraint in a certain number of trials, the cell was considered as not

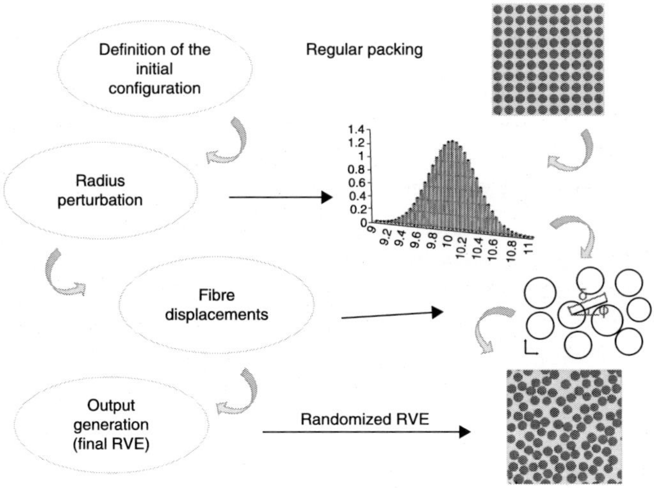

Figure 5.13 **Flow chart of the generation algorithm for the random RVEs**

generable using the parameters considered and a new cycle was started after reduction of the standard deviation.

After the creation of the initial cell, a further perturbation was enforced by randomly moving each fibre from its current position, following an iterative scheme. The displacement of each fibre was computed in a polar reference system centred on the centre of the fibre, defining the direction of the displacement according to an azimuth evaluated assuming a uniform distribution. The radial coordinate, which represents the modulus of the displacement, was defined as a fraction (randomly generated in [0, 1] following a uniform distribution) of the current minimum distance d_{min} between the moving fibre and all the other fibres constituting the current cell. A displacement was accepted if the constraint on the minimum distance d_{min}, as initially defined, was not violated. In the case of rejection of the displacement of a generic fibre, a successive fibre was considered until the end of the iteration. It should be noted that the movement of the fibres generated to enforce the symmetry condition was imposed by the parent fibre, considering always the requirement of no overlap. Taking into account the requirement that the unit cell should be periodically repeated, a symmetric geometry was enforced during the aforementioned operations by copying the fibres placed near the edge of the RVE in a manner corresponding to the next cell. The fibre displacement procedure was repeated for a predefined number of iterations N_i. An example of the evolution of a cell is shown in Figure 5.14. Here, N_i was

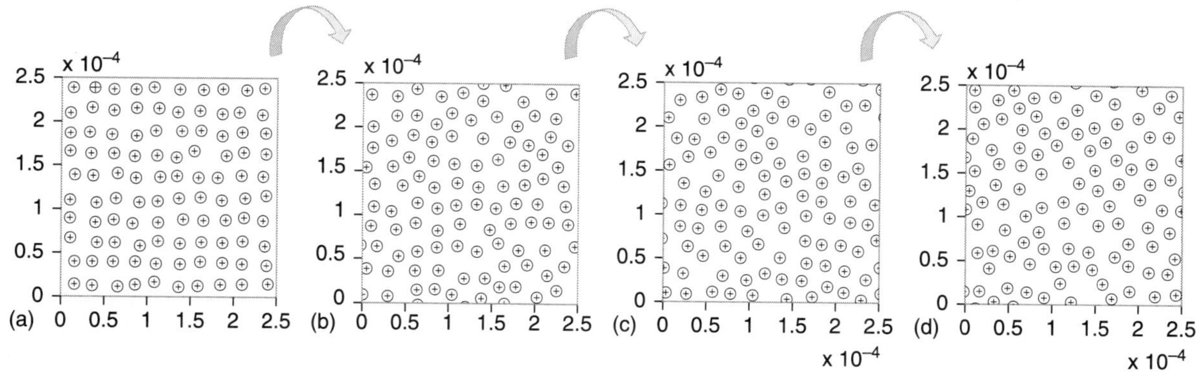

Figure 5.14 RVE perturbation: (a) 50, (b) 500, (c) 1000, (d) 5000 iterations

defined as 5000, following results provided by preliminary tests; a higher N_i did not lead to further disorder, since an autocompensating effect was observed. The other parameters, i.e. V_f, N_f, r and d_{min}, were taken as 0.6, 100, 10 µm and $0.1d_{min}$, respectively.

The RVEs obtained were statistically characterized by means of the indices described in Section 5.2.1. In Figure 5.15, some results representative of all the cases analysed are reported. In particular, the linear plot of the Ripley function and the pair distribution functions converge fairly well to the reference functions provided by the Poisson distribution (Figure 5.15(a) and (b)), evidencing the random nature of the fibre distribution as well as showing that the number of fibres used to create the RVE was enough to represent the material considered. This trend is not strongly supported by the Morishita number displayed (Figure 5.15(c)); however, this is related to the fact that this index is very suitable for point distributions, while in this case the natural presence of the fibre surrounding its centre conditions the calculation, particularly at higher volume fractions.

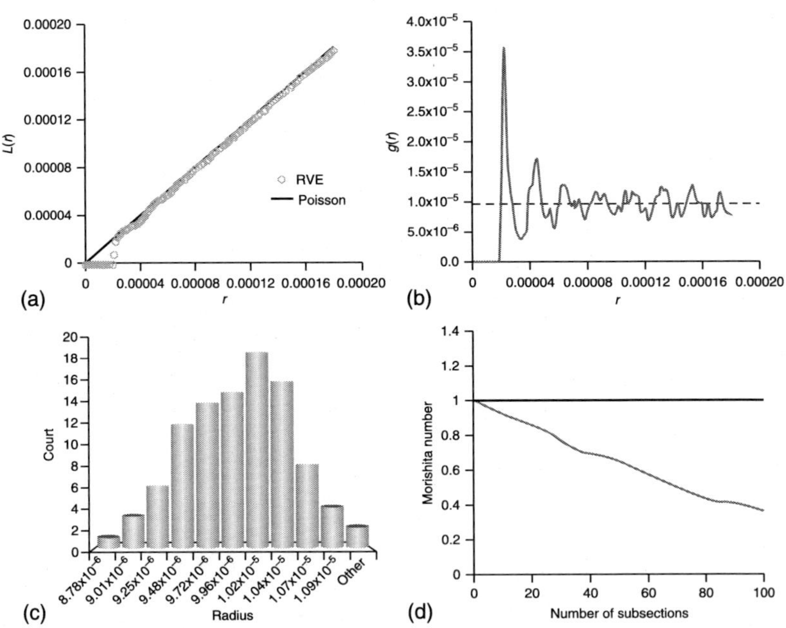

Figure 5.15 Statistical analysis of the RVEs: (a) Ripley *L* function; (b) pair distribution function; (c) histogram of radius distribution; (d) Morishita number

The homogenized transverse conductivity was calculated by means of the well-known Fourier model, applied to a finite element heat transfer model of the perturbed RVEs, built by means of routines implemented using the ANSYS Parametric Design Language (APDL). The quality of the mesh (in terms of element size and distribution) was defined by means of a sensitivity analysis, i.e. by comparing results provided by the simulation of a regular (square and hexagonal) arrangement with an ultrafine model of a single fibre until the results matched. The heat flow was obtained by solving the steady-state energy equation (excluding the generative term), closing the boundary value problem using the following conditions (see Figure 5.16):

- an imposed temperature (resulting in a 1 °C temperature difference ΔT) on two opposite boundaries (normal to the transverse direction) of the computational domain;
- periodic boundary conditions on other external surfaces;
- perfect thermal contact condition between fibres and resin.

The transverse conductivity K_z was inferred by replacing the numerical results in the following relation:

$$K_z = q_z \frac{\Delta T}{\Delta z}, \qquad [5.41]$$

where q_z is the average transverse heat flux through the inlet area and Δz is the RVE length.

Fibreglass and Shell EPON 9420/9470/537 epoxy resin were assumed as the reinforcing material and matrix, respectively. The relevant physical

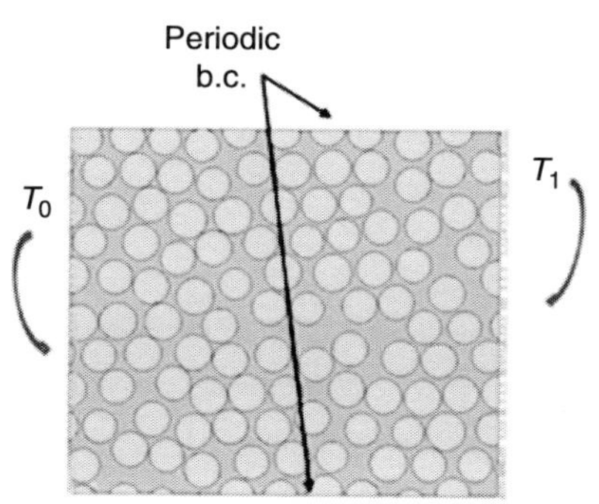

Figure 5.16 Micro-scale computational domain and boundary conditions

Table 5.4 **Physical properties and concentrations of materials**

Property	Fibreglass	Epoxy
ρ (kg m^{-3})	2560	1260
cp (J kg^{-1} K^{-1})	670	1255
Kx (W m^{-1} K^{-1})	11.0	0.21
Ky (W m^{-1} K^{-1})	1.04	0.21
Kz (W m^{-1} K^{-1})	1.04	0.21
Volume fraction	0.6	0.4

properties of each constituent are listed in Table 5.4, following the data reported in [62].

As far as the thermal flux is concerned, the numerical results clearly show that a stronger flux is induced in zones characterized by some alignment of the fibres along the applied temperature gradient, while an insulating effect is observable in resin-rich zones (Figure 5.17). As a consequence, a bridge effect is created, resulting in conductivity values higher than those obtained using homogenization by the simple rule of mixtures. A summary of the conductivity results, normalized with respect to the matrix conductivity K_m, obtained by running 12 different routines (six assuming a square initial packing and the other six assuming a hexagonal initial packing), is listed in Table 5.5 and compared with an analytical value provided by the simple rule of mixtures. As can be seen,

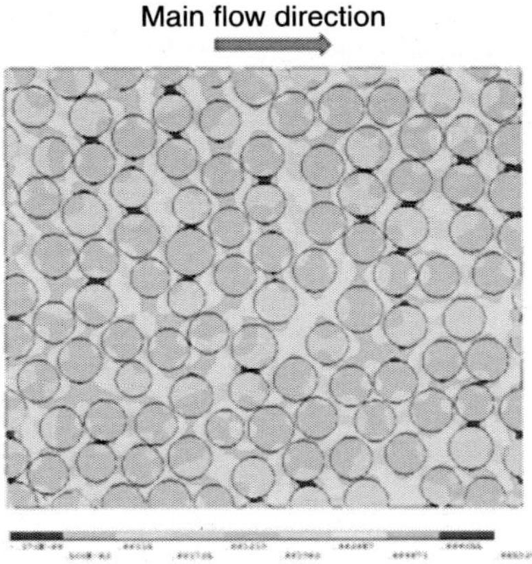

Figure 5.17 **Thermal flux along the transverse direction**

Table 5.5 **Normalized conductivity**

Routine	K_z/K_m	Mean	Standard deviation
1 (S)	2.252		
2 (S)	2.273		
3 (S)	2.191		
4 (S)	2.004	2.205	0.106
5 (S)	2.301		
6 (S)	2.208		
1 (H)	2.308		
2 (H)	2.338		
3 (H)	2.269	2.218	0.107
4 (H)	2.062		
5 (H)	2.192		
6 (H)	2.136		
Rule of mixtures	1.918	–	–

the analytical value represents an underestimate of the transverse conductivity, with the difference ranging between 4 and 21%. What is more, similar results were obtained assuming square and hexagonal packings as the initial configuration, highlighting the independence between the perturbed configuration and the initial regular configuration.

The effect of the stochastic variability of the radius and arrangement was investigated on the macro-scale by means of the one-dimensional thermochemical finite element model of the cure process described above. The heat generated by the resin reaction was included in the calculations, adopting the well-established *n*-order model to describe the species conversion [63]:

$$R_r = A_0 \exp\left(-\frac{\Delta E_i}{RT}\right)(1-\alpha)^n. \quad [5.42]$$

The parameters of the kinetics model adopted are reported in Table 5.6 [63].

Several different cure simulations were performed, adopting the minimum and maximum transverse conductivity computed on the micro-scale to define the material properties on the macro-scale, and these were compared with results obtained using the rule of mixtures.

Table 5.6 **Parameters of resin kinetics**

Parameter	Value
A_0 (s^{-1})	19.14×10^4
ΔE_0 (J mol^{-1})	60.5×10^3
n	1.69
H_{tr} (J kg^{-1})	323.7×10^3

The predicted results, i.e. the temperature and degree-of-cure profiles at the centre of the composite laminate undergoing processing (in order to make the effect of conductivity more visible), are shown and compared in Figure 5.18, including also the prescribed thermal cycle. As can be observed, the profiles obtained using the rule of mixtures and the minimum conductivity value provided by the FEM practically coincide, which is reasonable considering the negligible difference between the input data. However, more significant differences are detected in the heating and cooling stages when the maximum conductivity computed on the micro-scale is used, suggesting an opportunity for better consideration of stochastic effects in the homogenization of the material properties. The multi-scale non-deterministic approach adopted allows one to define a range of confidence describing the evolution of the process.

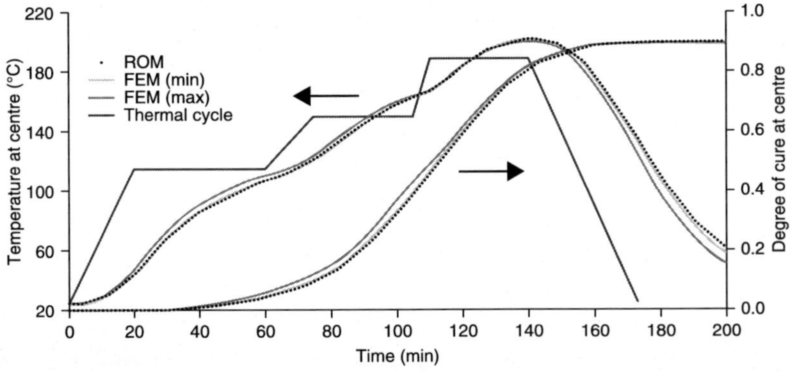

Figure 5.18 **Temperature and degree of cure at centre**

5.2.4 Pultrusion

Introduction

The pultrusion process has experienced remarkable growth within the composites industry, owing to its cost-effectiveness, automation and high quality of products, and, for the same reasons, it is currently receiving a great deal of attention in several key industrial sectors [64]. At the present time, most pultrusion processes are planned by means of trial and error approaches, consisting mainly in the start-up of a pultrusion line for a specific product. These procedures have made process know-how proprietary knowledge, creating a significant gap between industrial practice and an actual understanding of the phenomena involved.

Even if the process is conceptually quite simple, its analysis is a complex task, owing to the mutual interactions between the physical phenomena involved. In contrast to other composite-manufacturing processes, in pultrusion the steps performed to transform the raw materials (fibres and matrix) into the final product (a pultruded profile) are concentrated into a forming–heating–curing die, whose length is generally about one metre. Numerical and experimental investigations of different aspects of the pultrusion process have been reported in the literature, focusing mainly on issues related to heat transfer and curing [63, 65–74], the pressure distribution [75, 76] and the pulling force [77–83]. However, the proposed models often neglect the interactions between some of the phenomena involved, graphically schematized in Figure 5.19, on the

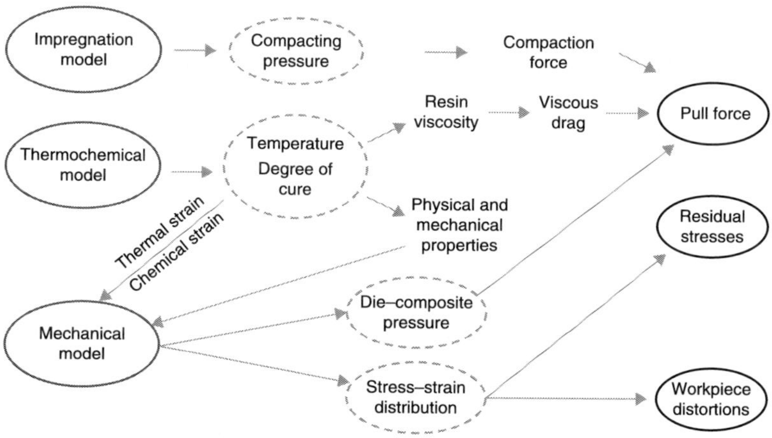

Figure 5.19 Multi-physics involved in the pultrusion process and related interactions [84]

basis of some simplifying assumptions. In what follows, some numerical models dealing with the analysis of impregnation, thermochemical aspects and the pull force are reported.

Impregnation analysis

In the conventional pultrusion process, reinforcing fibres are wetted out inside a resin bath before entering the heating die. After impregnation, the wetted fibres typically show an excess of resin with respect to the amount needed in the final product. As a consequence, in the tapered zone of the die (the inlet) the material being processed is compacted, resulting in a pressure increase with respect to atmospheric. Our impregnation model describes the pressure and velocity fields in the resin in the first part of the die, including the tapered or rounded zone and a portion of the straight part of the die. The velocity and pressure in the reinforcement-free zones of the domain are inferred by means of a conjunct solution of the well-known mass and momentum equations, which, under the hypothesis of incompressibility of the liquid resin and neglecting body forces, can be written as follows in a Cartesian coordinate system:

$$\nabla \cdot u = 0, \\ \mu \nabla^2 u = \nabla P, \qquad [5.43]$$

where u is the velocity vector and P is the liquid pressure. In particular, since the early part of the die is not heated in order to avoid premature resin gelation, the variations in the temperature and the degree of cure are negligible, and therefore the resin viscosity μ remains constant.

The reinforcing fibres are treated as a moving porous medium, in which the porosity and the permeability vary according to geometrical considerations, ensuring always the correct final fibre volume. The following modified Darcy model is solved in the porous region:

$$u = U - \frac{K}{\mu} \cdot \nabla P, \qquad [5.44]$$

where U represents the velocity vector of the porous medium. It should be noted that the component of the velocity vector U aligned with the pull direction is constant, and it is the only non-zero term in the straight portions of the domain, while other components need to be locally modified by considering the geometric configuration of the tapered zone [75, 84]. The FVM, as implemented in the commercial package ANSYS-CFX, was used here to solve the impregnation model, particularized to the case study reported in [76], simulating the pultrusion process of a unidirectional graphite/epoxy composite rod with a circular cross-section whose radius was 4.75 mm. The pultruded composite rod consisted of Shell EPON 9420/9470/537 resin and Hercules AS4–12K graphite fibres. The computational domain was restricted to the first 30 mm of the die, assuming that after this length any flow perturbations induced by the convergent section of the inlet vanish. The tapered inlet was modelled assuming a rounded shape with a length L_t and radius R_t equal to 6 and 6.35 mm, respectively [76]. The preform ratio, defined as the ratio between the cross-sectional areas of the impregnated material before and after compaction due to the tapered inlet, was assumed to be 1.44, neglecting shape variations of the pulled material. As a consequence, the wetted fibres approaching the inlet were modelled as a cylindrical porous medium with a radius equal to 5.7 mm.

The tow permeability was defined according to the Gebart model, which, assuming the x axis as the pull direction, yields

$$K_{xx} = \frac{8r_f^2}{c}\frac{(1-V_f)^3}{V_f^2},$$
$$K_{yy} = K_{zz} = C_1\left(\sqrt{\frac{V_{f}, \max}{V_f}}-1\right)r_f^2, \qquad [5.45]$$

where r_f (13 µm) is the fibre radius, $V_{f,max}$ is the maximum achievable fibre volume fraction, and C_1 and c are constants equal to 0.231 and 53, respectively [76]. As aforementioned, a constant viscosity was assumed, taking into account that generally no significant reaction is observed in the very early part of the die. The reference viscosity value was computed obtained according to Eq. 5.20 and the parameters listed in Table 5.7, considering the resin as fully uncured ($\alpha=0$) at a temperature equal to 38 °C. It should be noted, however, that the catalysed resin, before

Table 5.7 Parameters of the rheological model

Parameter	Value
μ_∞ (Pa s)	5.12×10^{-7}
$E\mu$ (J mol^{-1})	$3.76\times10^{+4}$
K	45.0
R (J K^{-1} mol^{-1})	8.314

entering the die, lay in the open bath for some time. During this period, a small amount of reaction cannot be excluded a priori. Even if the degree of cross-linking in the resin bath does not significantly affect the evolution of the solidification process, it could influence the viscosity for the impregnation and compaction analysis. This aspect was investigated by simulating the compaction process using three different viscosity values: 1.05 Pa s ($\alpha = 0$), 1.5 Pa s ($\alpha = 0.008$) [76] and 2.60 Pa s ($\alpha = 0.02$).

The die surfaces were modelled as rigid walls, defined with a no-slip condition. An inlet condition was imposed on the inlet surface corresponding to the preform, while an opening condition allowing the creation of a resin backflow was applied on the surrounding surface. In both cases, a zero relative pressure was imposed. The velocity of the material crossing the outlet section was assumed to be equal to the pull speed v_{pul}, taken as 5 mm/s [76]. The computed profiles of the pressure rise at the centreline of the processing material are shown in Figure 5.20 and compared with reference data [76].

The excellent agreement found confirms the validity of the numerical model implemented and its usability for process analysis. In all the cases simulated, an increase in pressure was detected before the intersection point, which was identified by a contact between the reinforced preform and the internal surface of the die. This pressure variation was due to the effect of the resin backflow (Figure 5.21), which prevents the free flow of

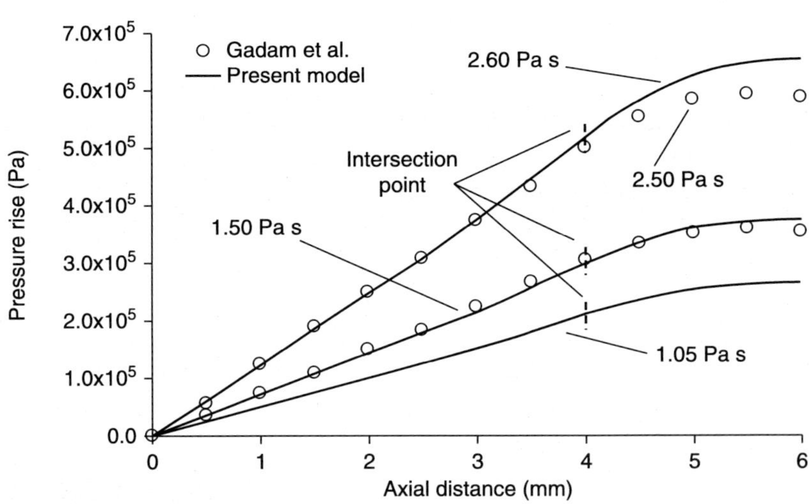

Figure 5.20 **Centreline pressure rise in the tapered region of the die**

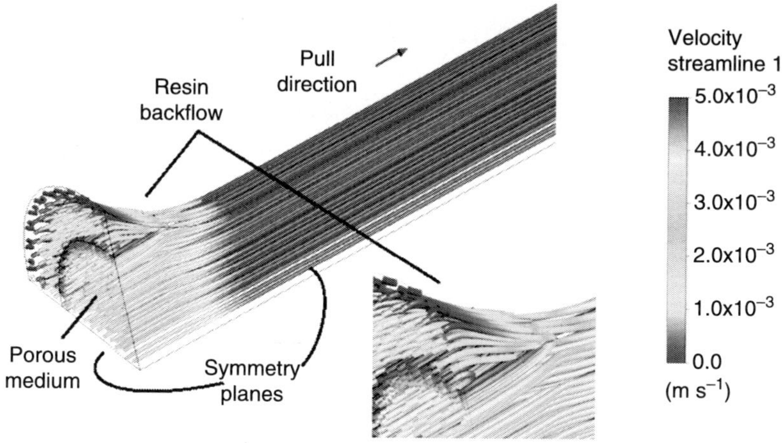

Figure 5.21 **Streamlines of resin flow in the tapered region of the die**

resin inside the preform towards the non-reinforced zones. Furthermore, more than half of the total pressure increase had already developed at the intersection point. It is also worth noting that, in Figure 5.21, at the very beginning of the straight portion of the die, the resin velocity converges to the pull speed imposed on the reinforcing fibres.

Thermochemical model (FDM and FEM)

In the straight portion of the die, heat provided by means of electrical heaters or hot oil activates the exothermic cure reaction. As a consequence, the material changes its state from a reactive liquid to a gel and then a vitrified solid [2, 64], following the TTT diagram [2, 8, 64] of the thermoset resin being processed. The thermochemical problem is a crucial issue for competitive production by pultrusion. Luckily, the pultrusion process is basically a steady-state continuous process. From a modelling point of view, the impact of this consideration is huge, since the time dependence is excluded from the governing equations and, even more appreciated, the computational domain is fixed in space.

Assuming constant properties and thermal equilibrium between the resin and the fibres (i.e. at each location in the composite material undergoing processing, all the constituents experience the same temperature), the whole temperature field can be established by solving a unique non-linear equation using lumped (homogenized) material properties [4–11, 16–18], which can be written as

$$\rho_c c_{p,c}\left(\frac{\partial T}{\partial t}+v_{\text{pul}}\frac{\partial T}{\partial x}\right)=k_{x,c}\frac{\partial^2 T}{\partial x^2}+k_{y,c}\frac{\partial^2 T}{\partial y^2}+k_{z,c}\frac{\partial^2 T}{\partial z^2}+V_r q, \qquad [5.46]$$

where T is the temperature; t is the time; ρ_c is the density; $c_{p,c}$ is the specific heat capacity; $k_{x,c}$, $k_{y,c}$ and $k_{z,c}$ are the thermal conductivities of the composite material along the x (pull), y and z directions, respectively; and V_r is the resin volume fraction. The source term q in Eq. 5.46 can be expressed as in Eq. 5.10.

The species equation is easily obtained from Eq. 5.18, considering that the transverse resin flow vanishes after the tapered section, where no reaction is observed in practice, and can therefore be neglected, yielding

$$\frac{\partial \alpha}{\partial t}+v_{\text{pul}}\frac{\partial \alpha}{\partial x}=R_r. \qquad [5.47]$$

Two solution schemes, based on a finite difference and a finite element scheme, were adopted to solve the thermochemical model. Both models take account of heat transfer due to heating platens, convective boundary conditions, a die cooler at the die entrance, the anisotropy of the composite material, mass transport effects and the heat generated by the exothermic cure reaction of the resin. Note that the time derivative in Eqs 5.46 and 5.47 are zero, since the process is stationary; however, they have been retained here for greater clarity in the finite difference model described below. As far as the finite difference model is concerned, taking into account that a second-order derivative can be written using a second-order central finite difference formulation, and considering the time derivative, one finally has seven dependent variables in each equation. A fast and efficient procedure can be obtained by decoupling the generative term due to the resin reaction, taking account of the values obtained in the previous integration step, as follows:

$$\rho_c c_{p,c}\left[\left(\frac{\partial T}{\partial t}+v_{\text{pul}}\frac{\partial T}{\partial x}\right)\right]^{n+1}=\left[k_{x,c}\frac{\partial^2 T}{\partial x^2}+k_{y,c}\frac{\partial^2 T}{\partial y^2}+k_{z,c}\frac{\partial^2 T}{\partial z^2}\right]^{n+1}+[V_r q]^n, \qquad [5.48]$$

$$\left[\frac{\partial \alpha}{\partial t}\right]^{n+1}=\left[R_r-v_{\text{pul}}\frac{\partial \alpha}{\partial x}\right]^n. \qquad [5.49]$$

In the conventional pultrusion process, heat is transferred to the composite material by means of a heated die. In most cases an appropriate temperature profile of the die is imposed by setting the control temperature of one or more strip (or cartridge) heaters. As a consequence, the heat transfer inside the die should be included in the analysis by solving the following equation:

$$\rho_d c_{p,d} \frac{\partial T}{\partial t} = k_d \frac{\partial^2 T}{\partial x^2} + k_d \frac{\partial^2 T}{\partial y^2} + k_d \frac{\partial^2 T}{\partial z^2}, \tag{5.50}$$

where ρ_d is the density, $c_{p,d}$ is the specific heat capacity and k_d is the thermal (isotropic) conductivity of the die material (steel).

The efficiency of the scheme can be improved by splitting the system of algebraic equations into a number of substeps equal to the number of dimensions considered. In this way, in each substep, only three terms, associated with a particular coordinate direction, are treated implicitly. As a consequence, the implicit terms can be grouped so as to be adjacent to the main diagonal and the solution can be obtained by the Thomas algorithm. The splitting technique adopted was the alternating direction implicit (ADI) method [85, 86]. The ADI scheme is characterized, for three-dimensional problems, by computational economy, spatially second-order accuracy but only conditional stability. The finite difference formulation of Eq. 5.48 can be written as follows:

$$\begin{aligned} T^{n+1}_{i,j,k} - T^{n}_{i,j,k} &= -C(T^{n+1}_{i,j,k} - T^{n+1}_{i-1,j,k}) + R_x(T^{n+1}_{i-1,j,k} - 2T^{n+1}_{i,j,k} + T^{n+1}_{i+1,j,k}) \\ &+ R_y(T^{n+1}_{i,j-1,k} - 2T^{n+1}_{i,j,k} + Ti,j+1,kn+1 + RzTi,j,k \\ &- 1n+1 - 2Ti,j,kn+1 + Ti,j,k+1n+1 + Qn, \end{aligned} \tag{5.51}$$

where

$$R_x = \frac{k_{x,c}\Delta t}{\rho_c c_{p,c} \Delta x^2}, R_y = \frac{k_{y,c}\Delta t}{\rho_c c_{p,c} \Delta y^2}, R_z = \frac{k_{z,c}\Delta t}{\rho_c c_{p,c} \Delta z^2}, C = \frac{\Delta t}{\Delta x} v_{pul}, Q = \frac{V_r q}{\rho_c c_{p,c}}. \tag{5.52}$$

Splitting Eq. 5.51 into the three directions of the coordinate system and assuming the time step for each substep to be $\Delta t/3$, the substeps associated with the x, y and z directions, respectively, can be written as follows:

$$\begin{aligned} &-(C + R_x)T^{n+1/3}_{i-1,j,k} + (1 + C + 2R_x)T^{n+1/3}_{i,j,k} - R_x T^{n+1/3}_{i+1,j,k} \\ &= (1 - R_y - 2R_z)T^{n}_{i,j,k} + R_y(T^{n}_{i,j-1,k} + Ti,j+1,kn \\ &+ RzTi,j,k-1n + Ti,j,k+1n + 13Q^n, \end{aligned} \tag{5.53}$$

$$\begin{aligned} &-R_y T^{n+2/3}_{i,j-1,k} + (1 + 2R_y)T^{n+2/3}_{i,j,k} - R_y T^{n+2/3}_{i,j+1,k} \\ &= (C + R_x)T^{n+1/3}_{i-1,j,k} + (1 - C - 2R_x)T^{n+1/3}_{i,j,k} + R_x T^{n+1/3}_{i+1,j,k} + R_z(T^{n}_{i,j,k-1} \\ &-2T^{n}_{i,j,k} + T^{n}_{i,j,k+1}) + 1/3 Q^n, \end{aligned} \tag{5.54}$$

$$
\begin{aligned}
&-R_z T_{i,j,k-1}^{n+1} + (1+2R_z) T_{i,j,k}^{n+1} - R_z T_{i,j,k+1}^{n+1} \\
&= (C+R_x) T_{i-1,j,k}^{n+1/3} - (C+2R_x) T_{i,j,k}^{n+1/3} + R_x T_{i+1,j,k}^{n+1/3} + (1-R_y) T_{i,j,k}^{n+2/3} \\
&\quad + R_y (T_{i,j-1,k}^{n+2/3} + T_{i,j+1,k}^{n+2/3}) + 1/3\, Q^n .
\end{aligned}
\qquad [5.55]
$$

A similar analysis can easily be done for Eq. 5.50 also. The degree of cure was evaluated by solving Eq. 5.49 using the upwind method, as follows:

$$
\alpha_{i,j,k}^{n+1} = [R_r]^n \, \Delta t + (1 - C)\alpha_{i,j,k}^{n} + C\alpha_{i-1,j,k}^{n}. \qquad [5.56]
$$

In the finite element model implemented, the steady-state condition yields the following modification of the energy balances in Eqs 5.46 and 5.50:

$$
\rho_c c_{p,c} \left(v_{\mathrm{pul}} \frac{\partial T}{\partial x} \right) = k_{x,c} \frac{\partial^2 T}{\partial x^2} + k_{y,c} \frac{\partial^2 T}{\partial y^2} + k_{z,c} \frac{\partial^2 T}{\partial z^2} + V_r q, \qquad [5.57]
$$

$$
\rho_d c_{p,d} \frac{\partial T}{\partial t} = k_d \frac{\partial^2 T}{\partial x^2} + k_d \frac{\partial^2 T}{\partial y^2} + k_d \frac{\partial^2 T}{\partial z^2}, \qquad [5.58]
$$

again decoupling the generative term from the energy equation. According to the procedure developed, the steady-state temperature field was initially calculated taking into account the imposed boundary conditions, assuming the degree of cure to be zero and the temperature of the composite to be the temperature of the resin bath at each node. At each point of the composite material, an overestimated value of the reaction rate could be evaluated using a kinetic model as in Eq. 5.13 and then the degree of cure was computed according to the following equation, obtained from Eq. 5.47 in the steady-state condition:

$$
v_{\mathrm{pul}} \frac{\partial \alpha}{\partial x} = R_r . \qquad [5.59]
$$

The values of the degree of cure obtained were then used to evaluate a more accurate value of the reaction rate, and the new values of the heat generation rate were applied to evaluate the temperature field in the second step of the iteration. The above procedure was repeated, using the results of each load step to evaluate the reaction rate, the degree of cure and the heat generation to be used in the next step, until the temperature difference obtained for each node between two consecutive iterations satisfied a previously defined convergence criterion.

The pultrusion of a workpiece with a C cross-section was simulated, according to the benchmark discussed in [79]. Only half of the model was considered, for symmetry reasons. Figure 5.22 shows the geometry

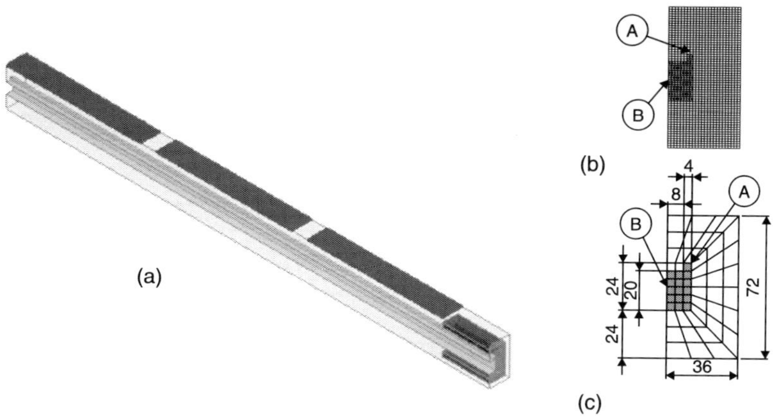

Figure 5.22 (a) Case study, and discretization of the cross-section: (b) FDM and (c) FEM

of the model (a) and its spatial discretization in the cross-section for the finite difference model (b) and finite element model (c). In the same figure, the points A and B in the section that were tracked regarding their temperature and degree-of-cure profiles are indicated.

The length L, width W and height H of the die were 915, 72 and 72 mm, respectively. The dimensions of the section of the composite are shown in Figure 5.22(c). The die was heated by six platens, whose dimensions were 255 mm (length) and 72 mm (width); three platens were placed on the top surface and the others on the bottom surface. The temperatures of the platens assumed are indicated in Table 5.8 (T_1, T_2 and T_3 refer to the heating platens on the top of the die, from the die entrance to the die exit, and T_4, T_5 and T_6 refer to the heating platens on the bottom of the die, in the same order). Consecutive platens were spaced by 30 mm and the empty zone was subjected to convective boundary conditions. The resin bath temperature was 45 °C. Between the die entrance and the first platen, over a distance of 90 mm, a water cooling channel was placed to avoid premature gelation of the resin. The temperature of the die cooler was assumed to be 50 °C. The pull speed

Table 5.8 Temperatures of the six heating platens (°C)

T_1	T_2	T_3	T_4	T_5	T_6
105.5	148.5	200	115	146.5	200

was 2.299 mm/s. The temperature of the room was assumed to be 30 °C and the convective coefficient to be 10 W/mm^2 K. Symmetric sections were modelled using adiabatic conditions.

The following initial and boundary conditions were used to solve the boundary value problem formulated above:

- the temperature and the degree of cure of the material in the cross-section at the die entrance were constrained to the resin bath temperature and zero, respectively;
- the cross-section of the composite at the die exit was modelled as adiabatic;
- adiabatic conditions were also imposed in symmetric sections;
- constant temperatures were imposed at appropriate nodes to simulate the heating platens (Table 5.8) and the die cooler;
- convective boundary conditions were imposed on the external die surfaces;
- it was assumed that the conductive heat flow at the interface between the die and the part was orthogonal to the die axis, taking into account the relatively low pull speed.

Shell EPON 9420/9470/537 epoxy resin and glass fibre were taken as the matrix and reinforcement. The physical properties and the kinetic model and its parameters can be found in Table 5.4, Eq. 5.43 and Table 5.6, respectively. The resin volume fraction was defined to be 0.361. The homogenized properties of the composite and the (chrome steel) die are listed in Table 5.9.

In Figure 5.23, the temperature profiles predicted by the finite difference and finite element models implemented are shown, evidencing the good agreement found. As can be seen, the temperature at point A is strongly

Table 5.9 Physical properties of materials

Property	Composite	Die
ρ (kg m^{-3})	2090.7	7833
c_p (J kg^{-1} K^{-1})	797.3	460
k_x (W m^{-1} K^{-1})	0.905	40
k_y (W m^{-1} K^{-1})	0.559	40
k_z (W m^{-1} K^{-1})	0.559	40

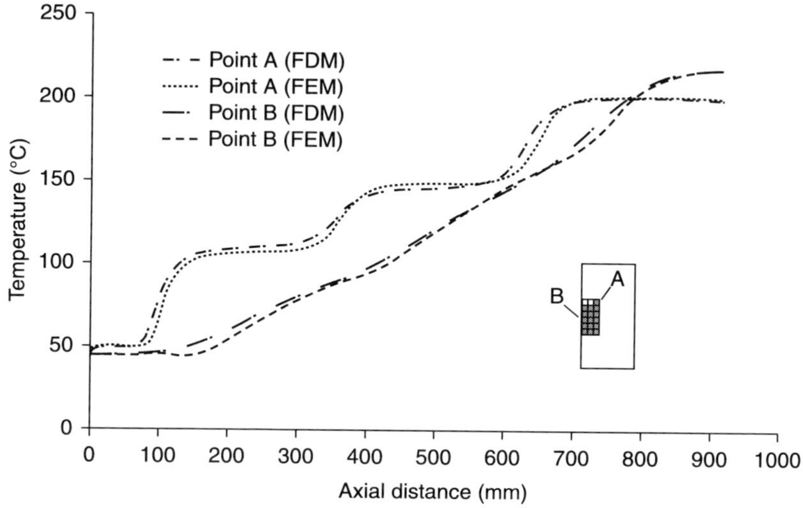

Figure 5.23 Temperature profiles in the pultrusion die

affected by the temperatures imposed on the heating platens, while at point B, the composite material experiences a sort of continuous heating until the exit section. The progressive, sharp increases in the control temperature prevail over the generative effect due to the resin reaction until the exit section is approached, when a temperature crossover is found (at approximately 780 mm from the die entrance). The maximum temperature of the composite was calculated as 215.85 °C by the FDM and 217.27 °C by the FEM; these values are both very close to the reference value of 217.72 °C [73].

The calculated degree of cure profiles at the test points A and B are depicted in Figure 5.24, together with reference data from [15]. The conversion profiles depicted show an earlier activation of the cure reaction at the composite surface due to a rapid temperature increase related to conductive heat transfer from the die wall (as also shown in Figure 5.23). As a consequence, the degree of cure at point A is initially higher than at the centre (point B). This trend is preserved over all of the die length, owing to the continuous increase in the platen temperature, which strongly promotes cross-linking of the monomers. The cure crossover is approached only in the die exit section of the domain, where the degree of cure values in A and B are very close. Considering the elevated temperature and the slope of the cure profiles at the die exit, a further increase in the degree of cure can reasonably be expected in the post-die region.

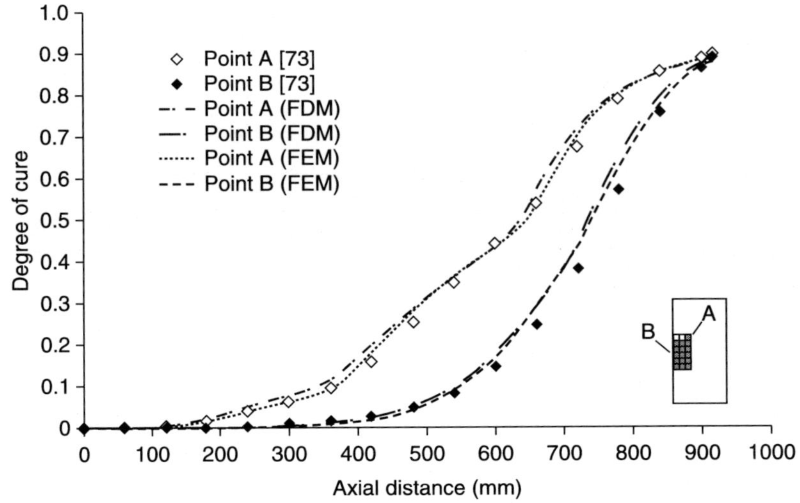

Figure 5.24 **Cure profiles in the pultrusion die and comparison with reference data [15]**

The mean degree of cure in the exit section of the composite workpiece was estimated as 0.888 (standard deviation 0.0036) by the FDM and 0.891 (standard deviation 0.0049) by the FEM, matching fairly well the reference value of 0.892 (standard deviation 0.0045) [73]. A remarkable accuracy was achieved using the proposed finite element method with respect to the finite difference model; the finite difference model seems to provide underestimates for all of the parameters considered. These differences are influenced by the spatial discretization in the model in the pull direction (the numbers of nodes in the pull direction were 101 and 1948 for the finite difference and finite element models, respectively) and in the cross-section. In the FEM simulation, the mesh was chosen considering the restrictions imposed by a condition related to the Péclet number. The relatively high Péclet number adopted in the FDM simulation indeed affects the accuracy of the solution. The induced error, however, can be considered acceptable, taking into account the reduced computation time. On the other hand, a drastic increase in the computational time for the FDM (with respect to the FEM) was required using the same value of the Péclet number. These features are expected, considering the conditional stability of the ADI scheme for three-dimensional problems.

Summarizing, the outcomes reported here show that the predicted results match quite well with each other and also with the available reference data. In particular, more accurate values for the peak temperature and the mean degree of cure were found using the finite element model, which provided an overestimate of the standard degree of cure. This shows that the finite difference and finite element numerical schemes adopted were stable and converged to the same solution. Both models could be used as effective tools to investigate the influence of process parameters, such as the temperature and location of the heating platens, the temperatures of the die cooler and the resin bath, and the pull speed, on the profiles and distributions of the temperature and degree of cure. What is more, parts characterized by different volume fractions, as well as by different sections, can be easily investigated. In this regard, finite element modelling appears to be more appropriate for modelling parts characterized by irregular or curved cross-sections, taking into account the great flexibility in the disposition of nodes in the cross-section of the composite and die.

Thermochemical model (FVM)

In this section, a finite volume approach to the solution of the thermochemical problem in pultrusion is presented. In particular, two models, namely a homogenized (or continuous) [81] and a non-homogenized (or porous) [84] model, were solved using the same numerical scheme. The basic assumption of the homogenized model is that of thermal equilibrium between different phases. In other words, at each location in the composite material undergoing processing, all of the constituents experience the same temperature. As a consequence, the whole temperature field is established by solving a unique non-linear equation using lumped material properties, coupled with a conversion equation governing the evolution of the species. The theoretical formulation of the model is identical to that described in the previous section.

Differently from the continuous model, the porous model treats the pultrusion process as the flow of a reactive liquid (the resin) through a moving porous medium (the reinforcement) inside a defined rigid boundary (the die cavity). It is a CFD-based non-thermal-equilibrium model that considers each component as a different entity on the macro-scale and admits a finite temperature difference (and heat transfer) between the reinforcement and the matrix. As a consequence, besides the continuity and momentum equations for the fluid phase, one energy balance equation for each component is needed. Assuming that the composite being processed is composed only of the reacting resin and the fibrous reinforcement, i.e. neglecting the effects of voids and porosity, the temperature field can be obtained by solving the following equations:

$$\varphi_{\rm f}\rho_{\rm f}C_{p,{\rm f}}\frac{\partial T_{\rm f}}{\partial t}+\rho_{\rm f}C_{p,{\rm f}}v_{\rm pul}\frac{\partial T_{\rm f}}{\partial x}=\varphi_{\rm f}\left(k_{x,{\rm f}}\frac{\partial^2 T_{\rm f}}{\partial x^2}+k_{y,{\rm f}}\frac{\partial^2 T_{\rm f}}{\partial y^2}+k_{z,{\rm f}}\frac{\partial^2 T_{\rm f}}{\partial z^2}\right)+Q_{\rm rf}, \quad [5.60]$$

$$\varphi\rho_{\rm r}C_{p,{\rm r}}\frac{\partial T_{\rm r}}{\partial t}+\rho_{\rm r}C_{p,{\rm r}}\left(u\frac{\partial T_{\rm r}}{\partial x}+v\frac{\partial T_{\rm r}}{\partial y}+w\frac{\partial T_{\rm r}}{\partial z}\right)$$
$$=\varphi\left(k_{\rm r}\frac{\partial^2 T_{\rm r}}{\partial x^2}+k_{\rm r}\frac{\partial^2 T_{\rm r}}{\partial y^2}+k_{\rm r}\frac{\partial^2 T_{\rm r}}{\partial z^2}\right)+\varphi q+Q_{\rm fr}, \quad [5.61]$$

where the subscripts 'r' and 'f' refer to the resin and fibre, respectively. In these equations, $\varphi = 1 - \varphi_{\rm f}$ represents the volume porosity of the medium (the ratio between the volume available for fluid flow and the total volume). Assuming the absence of voids, φ coincides with the resin volume fraction $V_{\rm r} = 1 - V_{\rm f}$. $Q_{\rm rf} = -Q_{\rm fr}$ is the interfacial heat transfer between the fluid and the solid, which depends on the temperature difference, the interfacial area density and the physical properties of the two phases. It should be borne in mind that in the porous model the degree of cure is treated as an additional scalar variable, with transport properties existing only in the fluid phase and varying according to a source term generated by the reaction rate. Similarly, the heat generation term q, as indicated in Eq. 5.10, is restricted to the reactive resin and the exothermic reaction affects the fibre temperature by means of conductive heat transfer. The ANSYS-CFX software package [87] was used to solve the porous thermal model, employing a finite volume numerical scheme.

The pultrusion process of a unidirectional graphite/epoxy composite rod with a circular cross-section (radius 4.75 mm) was simulated to compare the numerical outcomes provided by the proposed models, as well as to compare them with similar results discussed in the literature [70]. The inlet temperature was assumed to be equal to the resin bath temperature (38 °C), and the matrix material was assumed to be totally uncured ($\alpha = 0$) at the same cross-section. Considering that the composite material in the die exit section was still at an elevated temperature, it is reasonable to suppose that the cure reaction would proceed in the post-die region also, leading to some increase in the degree of cure. This aspect was included in the model by extending the length of the pultruded composite to the post-die region, ensuring that no further reaction would take place in the material. Relevant geometrical dimensions are indicated in Figure 5.25.

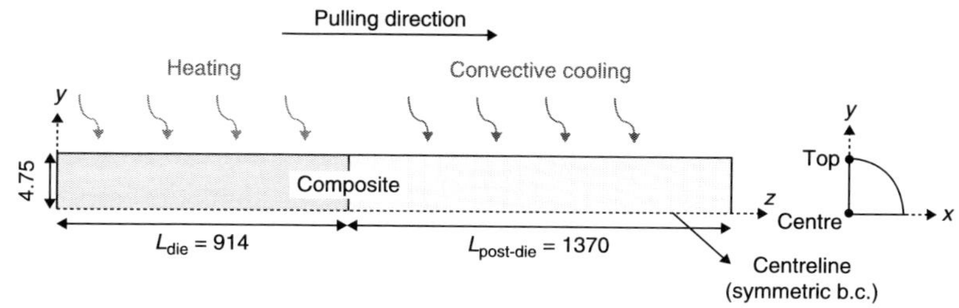

Figure 5.25 Schematic view of the pultrusion domain for the composite rod. All dimensions are in millimetres

In the post-die region, convective cooling to room temperature (27 °C) was imposed as a boundary condition on the external surface of the pultruded product. The dependence of the convective cooling coefficient on the surface temperature was defined using the well-known principle of heat transfer for a horizontal cylinder. The pull speed v_{pull} was defined as 5 mm/s [70]. It should be noted that in the simulations performed, the temperature distribution on the internal surface of the die was used to provide the required closure of the thermochemical problem described above, i.e. the die was not included in the calculation domain [70]. Only a quarter of the three-dimensional model was considered owing to symmetry and in order to reduce the computational effort.

The pultruded composite rod consisted of Shell EPON 9420/9470/537 resin and Hercules AS4-12K graphite fibres. The properties of the components and the kinetic parameters of the resin are listed in Table 5.10. The reaction model and kinetic parameters of the resin are specified in Eq. 5.43 and Table 5.6, respectively.

Table 5.10 Physical properties and concentrations of materials

Property	Graphite	Epoxy
ρ (kg m^{-3})	1790	1260
c_p (J kg^{-1} K^{-1})	712	1255
k_x (W m^{-1} K^{-1})	66	0.2
k_y (W m^{-1} K^{-1})	11.6	0.2
k_z (W m^{-1} K^{-1})	11.6	0.2
Volume fraction	0.6	0.4

The calculated temperature profile on the centreline is shown in Figure 5.26, together with the temperature profile imposed on the die wall, and the calculated degree of cure profile is shown in Figure 5.27. It can be seen that the predicted results match quite well with the experimental data available in [70]. This shows that the modelling approaches adopted are stable and converge to a reliable solution. The temperature in the centre of the composite rod becomes higher than the die wall temperature after approximately 390 mm from the die inlet owing to internal heat generation in the epoxy resin. At that point a peak in the reaction rate is obtained, inducing a sharp increase in the degree of cure. The maximum temperature of the composite is calculated as approximately 208 °C. What is more, in the post-die region, the degree of cure is increased slightly, which indicates that curing still takes place after the die exit, as also observed in [70]. The centreline degree of cure increases from 0.84 (at the die exit) to 0.87 (at the end of the process), while, at the surface, it varies from 0.80 to 0.83, indicating a global percentage increase of approximately 3.6%.

The degree of cure profiles depicted in Figure 5.27 highlight an earlier activation of the cure reaction at the surface of the composite due to a

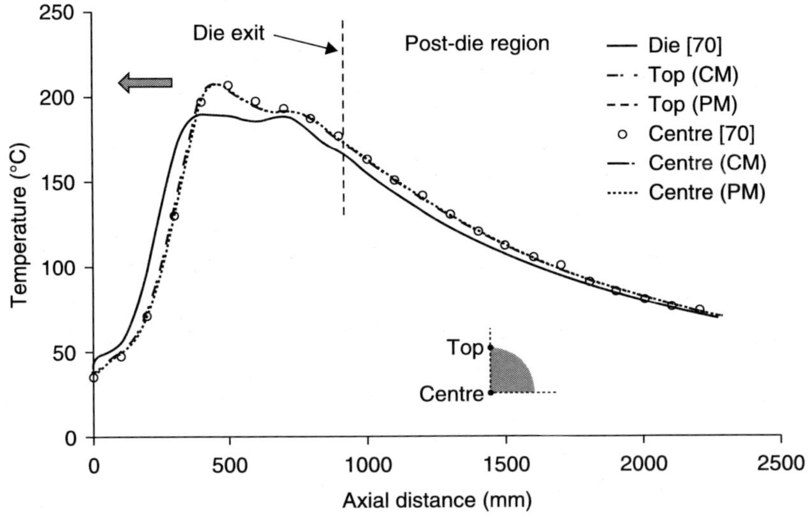

Figure 5.26 **Temperature and degree of cure profiles: comparison of the outcomes of the present calculations with the reference data [70] (CM, continuous model; PM, porous model)**

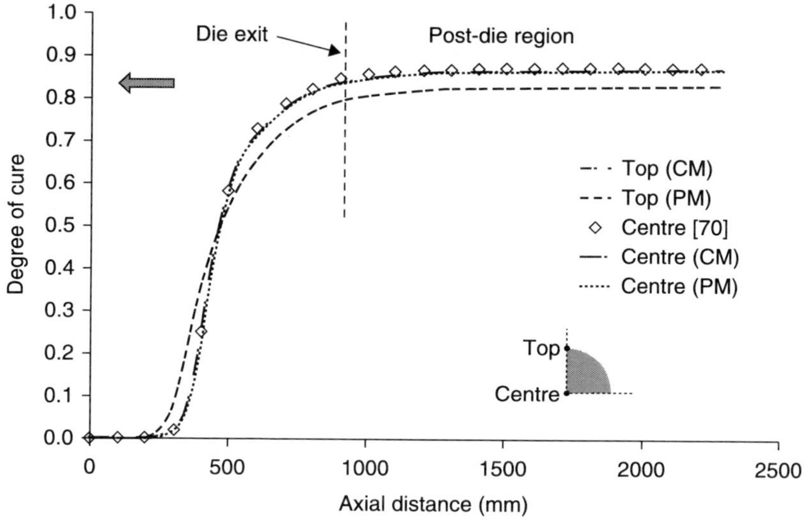

Figure 5.27 **Temperature and degree of cure profiles: comparison of the outcomes of the present calculations with the reference data [70]**

rapid temperature increase related to conductive heat transfer from the die wall. As a consequence, the degree of cure at the external radius is initially higher than at the centre. This trend varies after the activation of the reaction in the core of the material; indeed, the relatively low thermal conductivity of the resin prevents the heat generated at the centre from flowing towards the external zones, inducing a significant, localized temperature increase at the centre, which strongly promotes cross-linking of the monomers. It is worth noting that the cure crossover (the intersection between the degree of cure profiles at the centre and at the top) is reached at approximately $\alpha = 0.5$, which is well above the gel point ($\alpha = 0.265$) of the resin system considered, indicating a delay in the establishment of the desired inside–out solidification direction.

Further intriguing and very useful information, regarding for instance the pull force, can be obtained by post-processing the output of the thermochemical models illustrated. As mentioned before, most of the published work converges to the conclusion that the mechanical properties and quality of pultruded composites are strongly affected by the degree-of-cure distribution and the applied pulling force. The pulling force in pultrusion is attributable to several different contributions [2, 64, 77–83]: the collimation force F_{col}, the bulk compaction force F_{bulk}, the viscous drag F_{vis} and the frictional force F_{fric}, which are strictly related

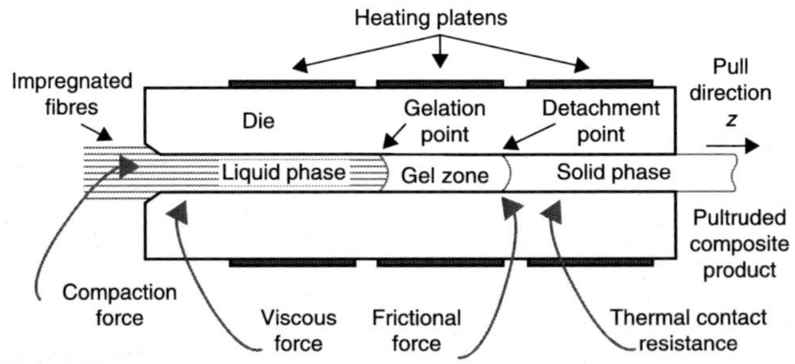

Figure 5.28 **Pulling force and phase changes**

to the geometrical features of the die–workpiece system and to the transitions of the resin from liquid to gel and then the solid state, as schematized in Figure 5.28.

The first contribution F_{col} is due to resistance between the creel and die inlet and is independent of the thermochemical aspects of the problem. F_{bulk} is related to the increase in the resin pressure typically observed in the initial tapered part of the die, when the resin is still in the liquid phase. This force acts along a direction normal to the die surface, and therefore, it does not affect the pulling force except at the tapered die entrance. Denoting the local resin pressure as p, the die taper angle as θ and the inlet surface as A_1, the bulk compaction term can be written as

$$F_{bulk} = \iint_{A_1} p \sin \theta \, dA. \qquad [5.62]$$

In the straight portion of the die, before the gel point, viscous drag occurs at the die wall. This resistance is attributable to the presence of a thin liquid layer between the travelling fibres and the stationary die surface. Thus, a plane Couette flow is induced, in which the reinforcing fibres can be represented as a moving plate translated at a constant pull speed and the die surface as a fixed plate. The viscous force can be written analytically as

$$F_{vis} = \frac{v_{pul}}{\lambda} \iint_{A_2} \mu(\alpha, T) \, dA, \qquad [5.63]$$

where λ is the thickness of the resin layer and A_2 is the surface affected by viscous effects, whose length is determined by the gel point.

After the gel point, the resin flow and the viscous effects are obviously inhibited and the composite is mechanically pulled through the die. Consequently, the interaction between the material undergoing processing and the die surface is characterized mainly by frictional effects. Denoting the friction coefficient as μ_f and the contact pressure as σ, this can be written as

$$F_{fric} = \int_{A_3} \mu_f \cdot \sigma \, dA, \qquad [5.64]$$

where A_3 is the die surface from the gel point to the detachment point. The magnitude of the contact pressure σ is considered to be affected by two contrasting effects: the transverse thermal expansion of the composite, due to the increase in temperature and pressure, and the chemical shrinkage of the resin, related to the cross-linking reaction [88]. The latter phenomenon leads to a progressive reduction in the size of the cross-section of the composite until it shrinks away from the internal wall of the die (the detachment point). Rigorously speaking, the separation of the material from the die cavity induces the formation of a thin (thermally insulating) air layer. As a consequence, a thermal contact resistance is interposed between the heated die and the material being processed. This aspect was not included in the simulations performed, since its effect was negligible [84]. The following results, summarized in Figure 5.29, were evaluated assuming the rheological behaviour of the resin as in Eq. 5.20 and Table 5.7. The evaluation of the geometrical variation of the workpiece and of the contact pressure was based on the semi-analytical approach detailed in [84].

The viscosity profiles depicted in Figure 5.29 show how the activation of the cure reaction (Figure 5.27) implies a sharp viscosity increase at

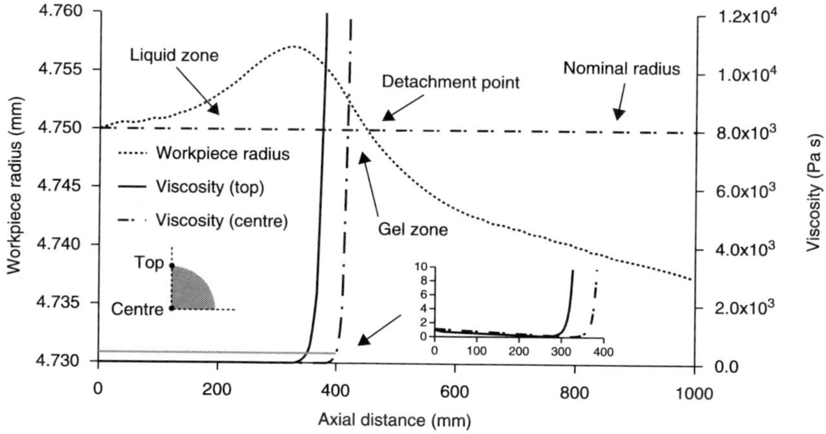

Figure 5.29 Viscosity profiles and virtual workpiece radius

gelation. This occurs earlier at the top surface, at a distance approximately equal to 360 mm from the die entrance, and separates the liquid zone (where the viscous drag acts) from the gel zone (dominated by the frictional resistance). The same viscosity trend is observed at the centre of the composite rod after approximately 405 mm from the die entrance. It should be also noted that in the first 200 mm from the inlet, the temperature increase (Figure 5.26) leads to a slight viscosity reduction before the beginning of cross-linking phenomena, as also highlighted in Figure 5.29. In the same figure, the workpiece radius is shown as a function of the axial distance. As highlighted by the numerical outcomes, in the liquid zone the thermal expansion prevails over the chemical shrinkage, leading to a virtual radius of the workpiece greater than the internal radius of the die. The detachment point, which is the intersection point between the virtual radius and the internal radius of the die during the shrinkage process, was found to be at approximately 545 mm from the inlet; as a consequence, the die length affected by the frictional effect is estimated to be approximately 185 mm. The workpiece radius in the exit section was found to be 4.742 mm, in good agreement with the value (4.739 mm) reported in [68]. For the case considered, the post-processing of the results provided by the impregnation and thermochemical models provided the following scenario: F_{bulk} = 4.9 N, F_{vis} = 313.7 N and F_{fric} = 184.1 N. The key role played by the viscous drag with respect to the frictional force can be related to the reduced die length affected by frictional phenomena and to the delayed development of the modulus of the resin (and the composite). The contribution due to compaction of the material was found to be not significant compared with the other contributions, being less than 1% of the total load.

5.2.5 Liquid composite moulding

In recent years, the increasing attention focused on the environmental and health impacts of polymer and composite manufacturing processes has promoted the development and industrial application of closed die systems, in order to avoid dangerous emissions of volatiles from the catalysed resin systems. This has been a driving factor related to the wider application of liquid composite moulding processes for the shaping of products and the impregnation of preforms, instead of the classical hand lay-up and spray lay-up processes [2]. What is more, remarkable advantages are offered by the RTM, VARTM and similar processes, as already highlighted in a previous chapter. The appropriate planning of LCM processes is, however, a very challenging task, since these processes are characterized by non-stationary multiphase flows in a three-dimensional porous domain with an anisotropic permeability. Furthermore, since the working fluid is catalysed and then reactive, variations in the temperature, degree of cure and, as a consequence, viscosity could be induced by the cure reaction, influencing the advance of the flow front. Finally, the elastic deformation of the fibre bundles due to the liquid pressure is not always negligible [3–7, 13–15, 23, 89–98].

Incomplete preform filling, non-uniform compaction, and variations in product thickness and the volume fractions of constituents are typical defects attributable to non-optimal process planning. So far, most liquid composite moulding processes have been designed, with regard to the definition of the positions of the inlets and the outlets, the working temperature and pressure, and the timing of each step, by adopting expensive trial and error procedures. Recently, a remarkable time (and cost) compression has been achieved in this context by employing advanced computational techniques to assist the analysis and optimization of the process, as shown well in the literature cited here. Although the existing simulation codes, developed thanks to the efforts of several research groups, have led to better understanding and more effective planning of LCM processes, a further open question is the development of online monitoring and control systems able to optimally handle unpredicted variations in the processing conditions. Different approaches to flow monitoring, based on image analysis, electrical signal processing, thermocouples, fibre optics, ultrasonic methods and dielectric measurements, have been proposed, compared and discussed in the literature [99–107], showing advantages but also limitations related to invasiveness, excessive noise, delay or tooling cost, depending on the specific choice. An automatic control system should have decision skills, based on a proper consideration of the different possible flow scenarios, in order to be effectively applied. Numerical simulations are excessively time-consuming and therefore often unsuitable for this purpose. However, they can be used to train faster models, based on ANNs for instance, to be recalled by an intelligent control system. In what follows, some applications of numerical models dealing with resin flow in single- and dual-scale porous media, including also some stochastic analysis, are reported. The finite volume method, as implemented in the ANSYS-CFX software package, was adopted to obtain the numerical solutions.

Single-scale isothermal flow model

This relatively simple problem, representing the first approach to the modelling of liquid composite moulding processes, is easily handled by several (generic or dedicated) commercial packages. The governing mass and Darcy equations, describing the (uncompressible) resin flow in an (undeformable) single-scale porous medium (such as a preform consisting of unidirectional long fibres), have already been indicated at the beginning of this chapter; however, they are briefly recalled here in the interests of clarity:

$$\nabla \cdot u = 0, \qquad [5.1]$$

$$u = -\frac{K}{\mu} \cdot \nabla P, \qquad [5.2]$$

which combined yield

$$\nabla \cdot \left(\frac{K}{\mu} \cdot \nabla P \right) = 0. \qquad [5.3]$$

Unidirectional flow simulations were performed in order to compare the numerical results with an analytical solution available in closed form, describing mono-dimensional flow through a porous medium with an isotropic permeability between two parallel boundaries [108], which can be written as

$$x_{\mathrm{f}} = \left(\frac{2KP_{\mathrm{d}}}{\mu\varphi} t \right)^{1/2}, \qquad [5.65]$$

where x_{f} is the distance of the resin flow front from the inlet, K is the isotropic permeability of the preform, P_{d} is the driving pressure, μ is the resin viscosity, φ is the preform porosity and t is the time elapsed since the beginning of the impregnation stage. Two different conditions were simulated, considering different epoxy resin systems, namely Araldite LY 1564 SP and NCS 236. In both cases, φ, K and P_{d} were assumed equal to 0.4, 1×10^{-10} m^2 and 0.9 atm, respectively. The properties of the resin systems considered are listed in Table 5.11.

Table 5.11 **Properties of the resin systems**

Property	NCS 236	Araldite LY 1564
ρ (kg m^{-3})	1050	1100
μ (Pa s)	0.182	0.246

The numerically evaluated preform filling at several different time instants after the beginning of introduction of the resin NCS 236 into the domain is shown in Figure 5.30. In Figure 5.31, the resin flow fronts for both resins are compared with the results provided by the analytical solution under the same processing conditions. As can be seen, the numerically and analytically

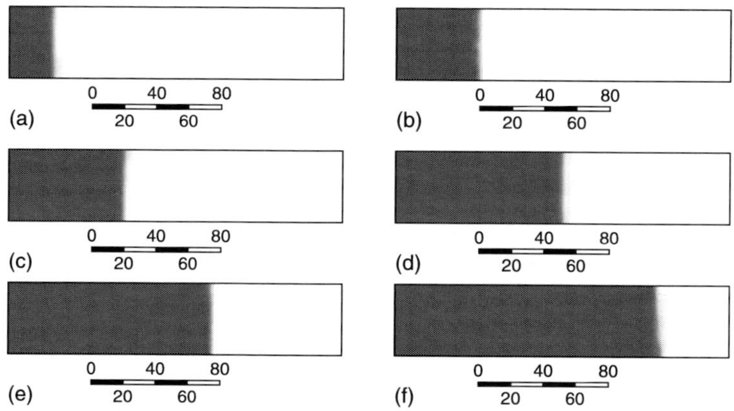

Figure 5.30 Flow front after (a) 10 s, (b) 30 s, (c) 60 s, (d) 120 s, (e) 180 s and (f) 300 s since the beginning of impregnation (length scales in millimetres)

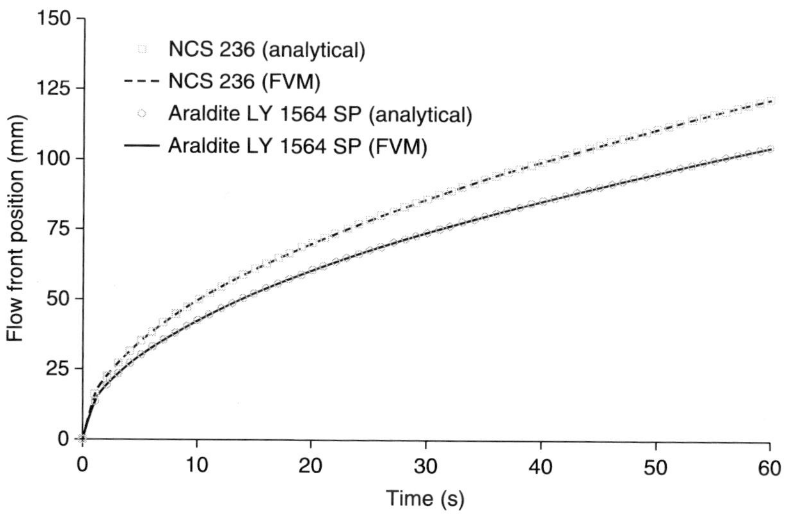

Figure 5.31 Resin flow front: numerical and analytical results

computed flow fronts practically overlap, attesting to the validity of the model for describing the preform impregnation stage under the isothermal assumption and neglecting the thermo-rheological effects of the cure reaction. It is easy to relate the differences in the positions of the flow fronts versus time to the viscosity values exhibited by the two resin systems considered. The high viscosity of the Araldite LY 1564 SP compared with the NCS 236 obviously implies slower impregnation of the preform.

A further case study will now be described, to better highlight the intriguing possibilities offered by numerical approaches to the simulation of preform wetting when closed-form solutions are not available. The geometry considered, shown in Figure 5.32(a), recalls a boat hull. Note that this should be interpreted as a purely demonstrative example and no actual design was made to define the general dimensions, indicated in Figure 5.32(a), of the part. A pressure difference of 0.87 atm was applied, modelling a distribution line (inlet) corresponding to the intersection of the geometry with its symmetry plane, and a vent line (outlet) at the upper edge of the hull. Taking into account the symmetry of the geometry and loads, the computational domain was restricted to one half of the model, which was discretized as depicted in Figure 5.32(b).

Araldite LY 1564 SP (see Table 5.11 for material properties) was adopted as the working fluid, and the porosity of the preform was 0.4. In Figure 5.33, the positions of the flow front at different time instants are shown, showing that full wetting of the preform was achieved 30 min after the beginning of the process.

Micro-scale stochastic analysis

As far as preform impregnation is concerned, a key parameter that needs to be accurately assessed to perform a reliable process simulation

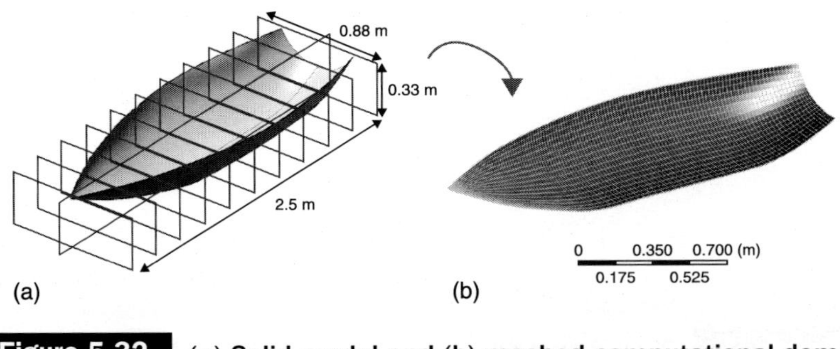

Figure 5.32 (a) Solid model and (b) meshed computational domain

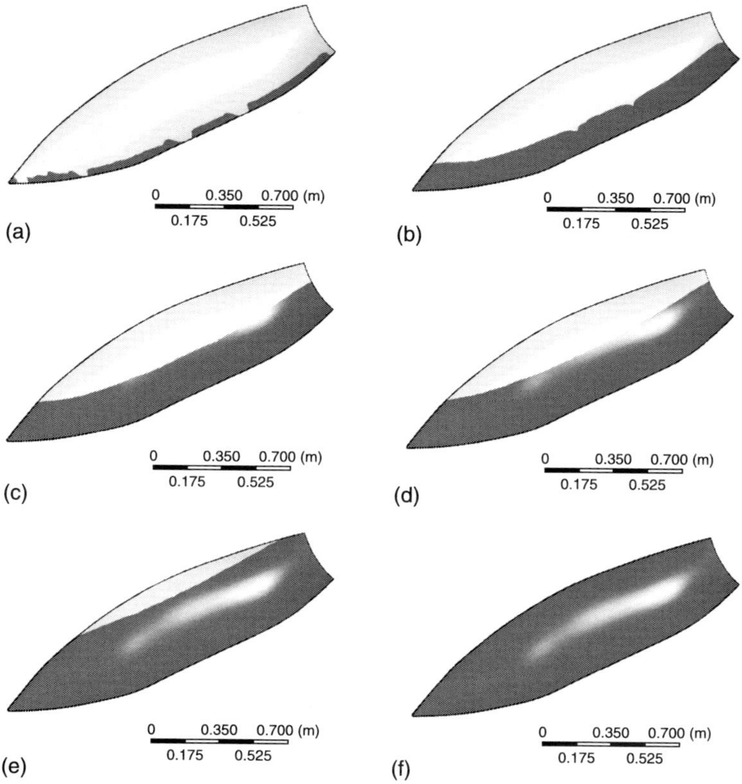

Figure 5.33 Resin front after (a) 15 s, (b) 3 min, (c) 6 min, (d) 9 min, (e) 18 min and (f) 30 min

is the permeability tensor. Interesting experimental techniques have been proposed and effectively applied to measure permeability in the recent past. Moreover, several models simulating the material behaviour on the micro-scale under the assumption of a regular fibre distribution have been implemented in the literature, with the aim of evaluating the permeability of the fibre tows to be used in meso- and macro-scale flow simulations [36, 37, 109]. The hypothesis of perfect fibre packing and the absence of dimensional variations between fibres, however, is in strong contrast to the experimental observations. The opportunity for proper consideration of the geometric disorder of the fibre distribution in RVEs has been remarked on in [39–45], as already stated. Following these previous studies, the influence of the stochastic variability of the fibre packing on the tow properties was numerically investigated, using a micro-scale model for the evaluation of the axial and transverse components of the permeability tensor of the fibre tows, under a non-deterministic assumption. The algorithm implemented for the generation and statistical characterization of the random RVEs was the same as that employed for the analysis of the thermal conductivity presented in Section 5.2.2. The values of the transverse and axial permeability were predicted by means of CFD simulations of the resin flow through the fibres in a computational domain defined by the perturbed RVEs. The continuity and momentum equations solved were of the form

$$\nabla \cdot u = 0 \qquad [5.66]$$

and

$$\rho(u \cdot \nabla u) = \mu \nabla^2 u - \nabla P, \qquad [5.67]$$

respectively, assuming steady-state flow (saturated permeability calculation) and a Newtonian, incompressible fluid, and neglecting body forces. The closure of the boundary value problem was defined according to the following boundary conditions (shown graphically in Figure 5.34):

- imposed pressures on the inlet (P_0) and outlet (P_1) surfaces of the computational domain;
- symmetry conditions on all other external surfaces (apart from the fibre surfaces);
- no-slip conditions (zero velocity) on the fibre surfaces.

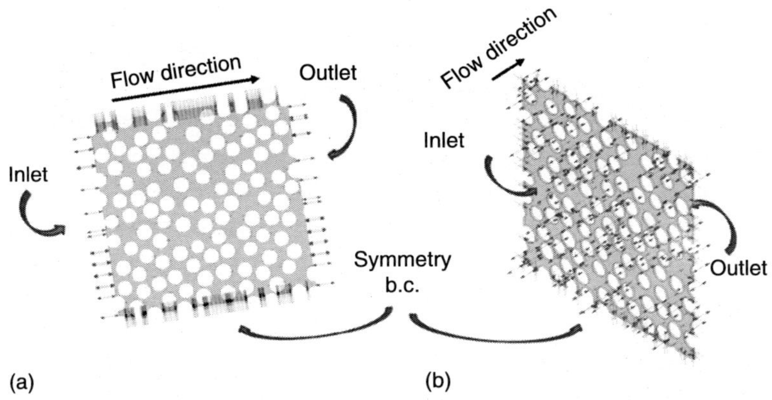

Figure 5.34 Computational domain and boundary conditions: (a) calculation of transverse permeability; (b) calculation of longitudinal (axial) permeability

The values of the transverse and longitudinal permeability K were inferred by means of the well-known Darcy model, by replacing the numerical results in the following relation:

$$q_x = -\frac{K}{\mu}\frac{\partial p}{\partial x}, \qquad [5.68]$$

where q_x is the flow rate (considering x as the direction of the main flow in a generic simulation) through the inlet area. The proposed procedure was validated in [110] by comparison with the analytical Gebart model [108], considering regular distributions of the fibres. Some permeability results provided by the computational simulations performed are reported graphically in Figures 5.35 and 5.36, relating to the transverse and axial (longitudinal) flows, respectively. For each fibre volume fraction, six routines (creation of the RVEs, model meshing and solution) were run. The quality of the mesh (in terms of element size and distribution) was defined by comparing the outcomes provided by an ultrafine model of a single-fibre (square or hexagonal) cell with a corresponding RVE including the same number of fibres (100) as in the disordered cells until convergence of the permeability results.

As can be seen, a certain scatter can be observed in the permeability values when the RVE considered is characterized by a random distribution of reinforcing fibres. In particular, a general reduction in the transverse permeability is found with respect to the initial regular cell, corresponding to lower volume fractions. The permeability variations can be related to

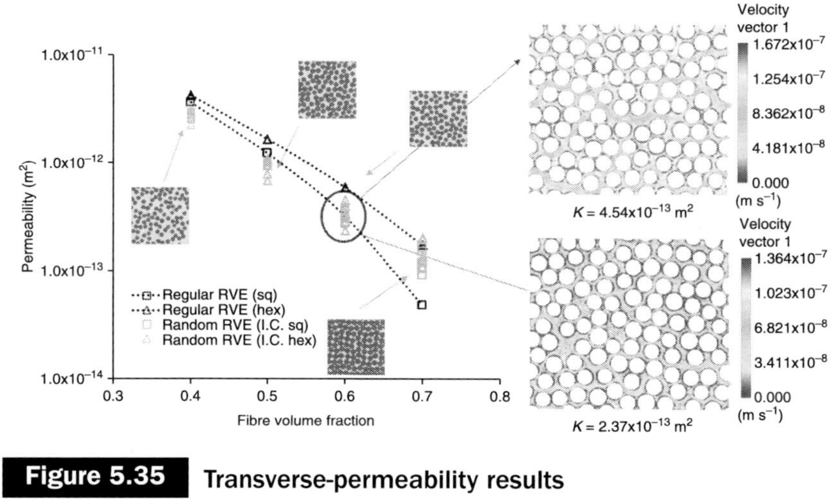

Figure 5.35 Transverse-permeability results

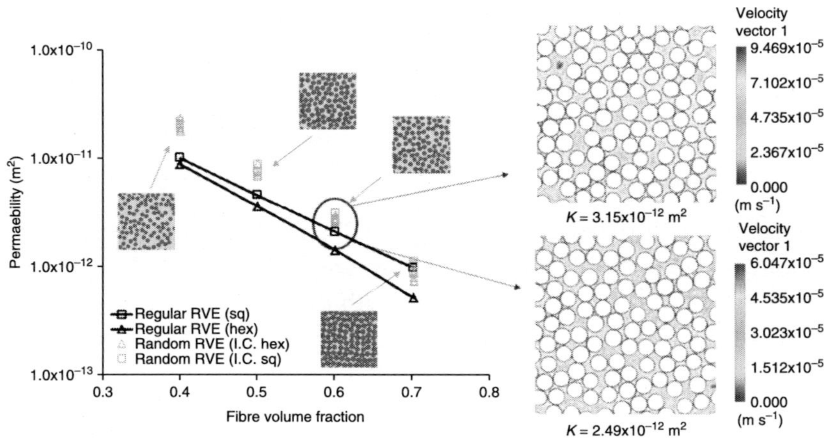

Figure 5.36 **Axial-permeability results**

two contrasting effects induced by the fibre rearrangement: a channel effect, due to the alignment of some fibres in the direction of the applied pressure gradient and leading to a permeability increase (Figure 5.35, top right), and a blocking effect, attributable to an opposite fibre disposition, increasing the tortuosity of the flow and reducing the homogenized permeability (Figure 5.35, bottom right). The remarkable variations (up to 100% between the minimum and maximum values, assuming the same fibre volume fraction) that are attributable to channel or blocking effects could be a plausible explanation for the deviations in the published experimental data. As far as the longitudinal permeability is concerned, the numerical results clearly show that the induced disorder and the randomization of the fibre distribution lead to the formation of wider empty zones characterized by higher flow rates (Figure 5.36). As a consequence, higher permeabilities are observed with respect to regular RVEs. This trend appears more pronounced at low volume fractions.

Single-scale non-isothermal reactive flow model

In the previous section, computational results obtained by considering an isothermal flow model were discussed. The effects of heat transfer and the cure reaction on the resin front advancement were neglected and a constant viscosity value was assumed. However, external heating sources and the exothermic cross-linking reaction can induce local temperature increases and, consequently, a decrease in the resin viscosity. This is very important, since it makes the resin flow through the preform easier, reducing the filling time. On the other hand, the temperature increase accelerates the cure reaction, resulting in a sharp viscosity increase, which can slow down or even stop the progression of the resin flow. Most impregnation models rely on the fundamental hypothesis of isothermal conditions and constant viscosity. As a consequence, their effective use in process planning and optimization, in particular when dealing with large parts, needs to be carefully checked. Moreover, in some liquid composite moulding processes, characterized by a post-cure stage performed in an autoclave or using heated dies, the distribution of the degree of cure at the end of the filling stage must be properly accounted for in order to design the cure cycle optimally. The influence of thermochemical phenomena on the impregnation stage is well highlighted in recent papers [15, 16, 23, 97, 111].

The impregnation model (Eq. 5.3) needs to be coupled with a heat transfer model, including also a transport variable, i.e. the degree of cure, to evaluate the local heat generation rate. In other words, assuming thermal equilibrium between the moving fluid and the stationary porous medium, the temperature field can be obtained by solving the following energy equation:

$$\rho_{\mathrm{eff}} c_{p,\mathrm{eff}} \frac{\partial T}{\partial t} + \rho_{\mathrm{r}} c_{p,\mathrm{r}} u_{\mathrm{r}} \cdot \nabla T = \varphi k_{\mathrm{eff}} \cdot \nabla^2 T + \varphi q, \qquad [5.69]$$

where T is the temperature, t is the time, ρ is the density, c_p is the specific heat capacity, k is the thermal conductivity tensor, u is the velocity vector and φ is the fluid volume fraction. The subscripts ‘r’ and ‘eff’ refer to the resin and to effective properties, respectively. However, considering that the generative term (Eq. 5.10) is also dependent on the degree of cure, a simultaneous solution of the transport equation (Eq. 5.18), including the source term given by the reaction rate, is needed. The computed temperature and degree-of-cure distributions must then be used to update the viscosity field, affecting the solution of Eq. 5.3.

An example of some simulations showing the effects of heat transfer (due to an external heating source at constant temperature as well as internal heat generation) and of resin conversion on the resin viscosity and, consequently, on the impregnation stage will now be reported. Three different simulations (S_1, S_2 and S_3) of the filling of a unidirectional preform were run, taking the resin temperature at the inlet as 25 °C, the heat transfer coefficient between the domain and the surrounding space as 30 W/m^2 K, and the external temperature T_{ext} as 25, 50 and 100 °C. The permeability of the preform was 5.92×10^{-11} m^2 (60 darcy), and the applied driving pressure was 0.9 atm. The resin system considered was

Table 5.12 Physical properties and concentrations of the resin

Property	Value
ρ (kg m^{-3})	1200
c_p (J kg^{-1} K^{-1})	700
k (W m^{-1} K^{-1})	0.5
Volume fraction	0.4

SI-ZG-5A epoxy resin [112]. Other relevant simulation parameters are listed in Table 5.12.

The following modification of the autocatalytic model in Eq. 5.16, including the diffusive term, was defined to describe the cure reaction:

$$\frac{d\alpha}{dt} = \frac{1}{1 + e^{C[\alpha - (\alpha_{co} + \alpha_{cT} T)]}} \cdot A\, e^{\left(-\frac{E_a}{RT}\right)} \cdot \alpha^m (1-\alpha)^n, \qquad [5.70]$$

particularized according to the parameters reported in Table 5.13 [112]. The rheological behaviour of the resin system was assumed to follow the Castro–Macosko model presented in Eq. 5.19, using the parameters [22] listed in Table 5.14 [112].

In Figure 5.37, the numerically computed resin fronts and temperature and degree-of-cure distributions provided by S_1 are shown. Frozen images are shown for two time instants, at 150 and 300 s after the beginning of the impregnation stage. Note that each image shows the distribution of the values in some planar sections of the domain, defined to have the shape of a block. As can be seen, for an external temperature of 25 °C, a relatively slow evolution of the flow front, covering a distance

Table 5.13 Parameters of the kinetic model

Parameter	Value
A (s^{-1})	3.03×10^5
E_a (J mol^{-1})	62×10^3
m	0.377
n	0.295
C	16.5
α_{c0}	0.65
α_{cT} (K^{-1})	-1.5×10^{-5}

Table 5.14 **Parameters of the rheological model**

Parameter	Value
A_μ (Pa s)	9.6×10^{-11}
E_μ (J mol^{-1})	5.48×10^4
A	3
B	8
α_g	0.6

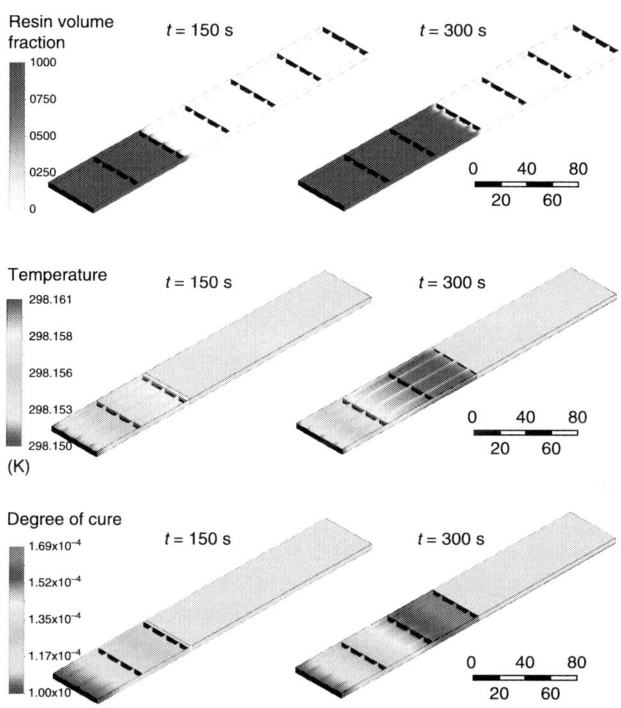

Figure 5.37 **Flow front, and temperature and degree of cure distributions at different time instants for T_{ext} = 25 °C. (Length scales in millimetres)**

approximately equal to 105 mm at 150 s and 165 mm at 300 s, is computed. A negligible temperature increase (about 0.01 °C) is found at 300 s, consistent with the fact that the temperature of the resin entering the domain is equal to room temperature, and therefore no external–internal heat fluxes are established. What is more, considering the reaction kinetics and parameters outlined in Eq. 5.70 and Table 5.13, no substantial reaction is activated, justifying the absence of effects due to any released heat. Under these operating conditions, the isothermal and non-isothermal reactive models practically coincide.

Some thermal effects are evident in Figures 5.38 and 5.39, which depict the position of the resin fronts, the temperature distributions and the degree of conversion calculated in S_2 and S_3, respectively. As can be seen, the higher the external temperature with respect to the temperature of the resin at the inlet, the higher the global temperature experienced by the material being processed, owing to the increasing energy stored in the domain. A very weak activation of the resin reaction, but more appreciable than in the first case analysed, is predicted in both cases at a time of 300 s; however, it is not actually sufficient to reverse the thermal effect on the resin viscosity. Owing to the viscosity reduction (Figure 5.40), a faster progression of the resin front is computed; indeed, for T_{ext} = 50 °C, the

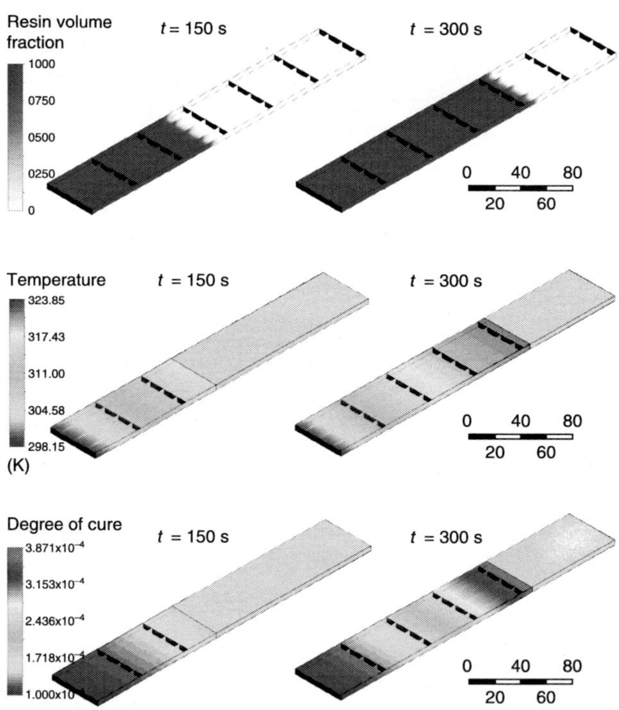

Figure 5.38 **Flow front, and temperature and degree of cure distributions at different time instants for T_{ext} = 50 °C. (Length scales in millimetres)**

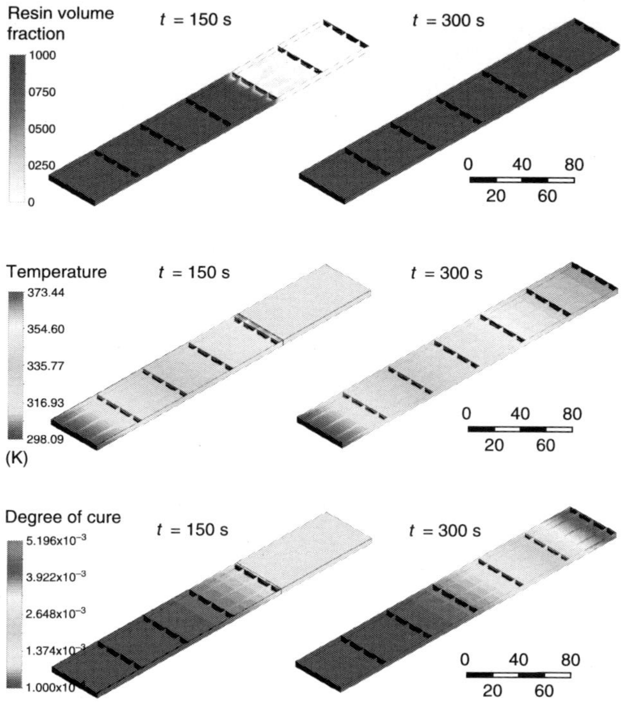

Figure 5.39 **Flow front, and temperature and degree of cure distributions at different time instants for T_{ext} = 100 °C. (Length scales in millimetres)**

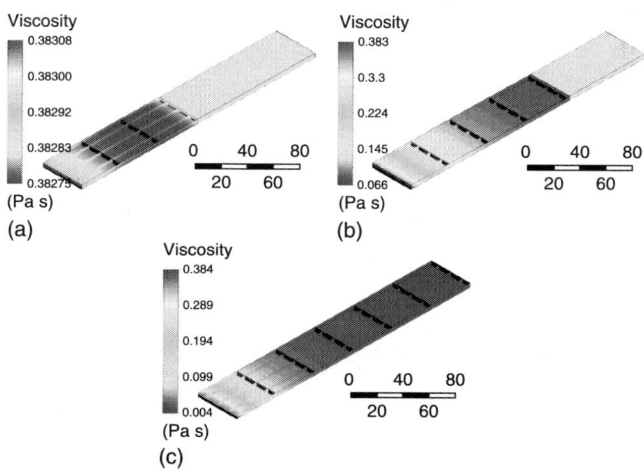

Figure 5.40 **Viscosity distributions at t = 300 s for T_{ext} = (a) 25 °C, (b) 50 °C and (c) 100 °C. (Length scales in millimetres)**

distance from the inlet was found to be approximately equal to 132 mm at 150 s and 211 mm at 300 s. This trend is even more marked for T_{ext} = 100 °C; in this case the whole preform, whose length was 300 mm, was impregnated in about 256 s.

The scenarios depicted can be explained by considering the slow reaction of the modelled resin system at temperatures not significantly greater than 100 °C. Indeed, according to the rheological model defined here, only the temperature effect is perceived by the fluid, resulting in a significant decrease in the global viscosity. The effect of external heating of the resin, in order to enhance the resin flow through the preform, has been also studied in [113]. The degrees of cure achieved were quite far from affecting the viscosity. It should be remarked, however, that a totally different evolution of the flow front is to be expected when highly reactive resins are simulated (or processed), in particular in the case of a low gel point. In these cases, the fast cross-linking of the resin monomers reasonably implies slower flows. As a consequence, the impregnation strategy, with regard to the inlet and vent positions, will be strongly conditioned by the reduced time available for complete wetting and saturation of the preform.

Figure 5.41 summarizes the flow front position with respect to the impregnation time provided by the simulations performed, and also compares the numerical outcomes with analytical results obtained from

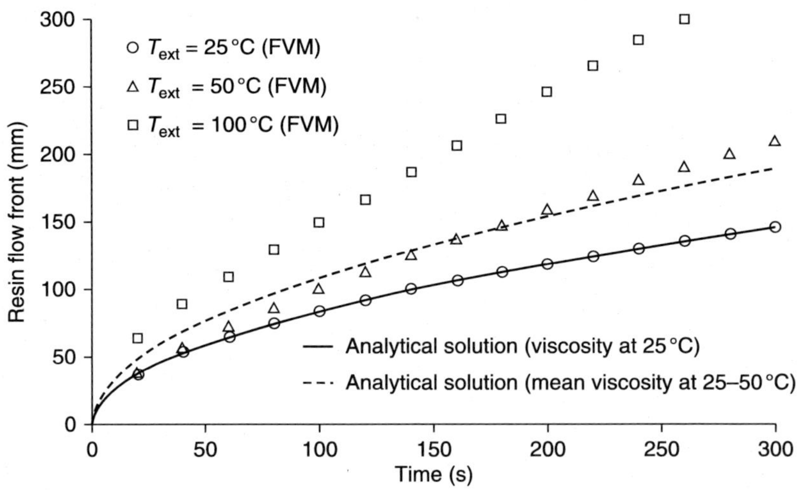

Figure 5.41 Numerical and analytically computed (using the mean viscosity value) resin flow fronts

Eq. 5.65, assuming the viscosity of the resin to be that of the uncured resin at 25 °C and the average of the theoretical values at 25 °C and 50 °C. The good overlap between the analytical and numerical curves at 25 °C confirms further that, in the absence of significant heating or activation of the cure reaction, isothermal models can be applied effectively, reducing the number of equations to be solved and, consequently, the computational time. On the other hand, the analytical and isothermal models fail when thermal factors or species conversion influences the viscosity distribution. A slight deviation can be observed when the numerical curve at T_{ext} = 50 °C is compared with the analytical curve (assuming the averaged viscosity value); however, the disagreement becomes unacceptable for T_{ext} = 100 °C, highlighting the appreciable capabilities of the numerical simulation. It should also be noted, however, that the solution of the model presented here requires a significant computational time (hours) even for small domains, and can reach some days for larger and more complex domains. As a consequence, even if some cases can be reasonably simulated at the process-planning stage, these models appear not to be compatible with the requirements of online control systems. For this purpose, meta-models based on artificial neural networks could be more effective. In this framework, the use of computational simulation for network training could provide intriguing benefits, in terms of time and cost compression.

Dual-scale isothermal flow model

Most of the commonly used reinforcing layers and textiles consist of complex arrangements of groups of fibres, called tows. These architectures are characterized by different levels of non-homogeneity depending on the specific observation scale, and are therefore referred to as dual-scale porous media. This term also implicitly indicates some differences in the permeability and porosity values inside the tows (where the fibres are strongly packed, reaching high volume fractions) and outside the tows. In this case it is appropriate to distinguish the macro-flow of the resin through the empty (inter-tow) spaces of the preform from the micro-flow inside each tow, where the latter leads to saturation and the removal of micro-voids.

A flow-monitoring system based on dielectric shielded sensors was employed in [107], where a unidirectional non-reactive flow through a dual-scale glass fibre preform was considered, as shown in Figure 5.42(a). The effect of saturation and the difference between the macro- and micro-flows are clearly indicated by the variations in the slopes of the curves

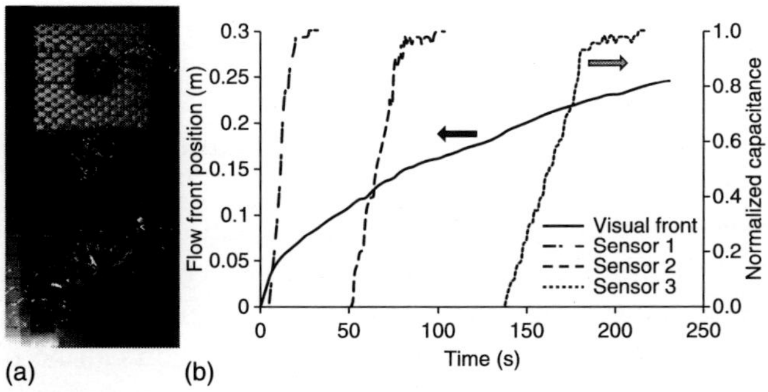

Figure 5.42 (a) Experimental set-up and (b) results from dielectric monitoring of flow through a dual-scale porous medium [107]

describing the normalized capacitance acquired by each sensor. In the first part of the curves, the capacitance varies with the flow front position in a reasonably linear fashion; however, at a certain time, roughly corresponding to the instant when the visually monitored (unsaturated) front reaches the end of the sensing plate, a slower capacitance increase (due to progressive saturation of the tows between the plates of the sensor) is measured, in agreement with the high resistance to the intra-tow flow compared with the inter-tow flow.

The modelling of the flow in dual-scale porous media is the subject of a vast literature [3–5, 7, 13–15, 89, 93, 96, 98, 114–116]. As aforementioned, two distinct approaches have been proposed to simulate the flow of resin through a dual-scale preform: the first is based on the inclusion of one-dimensional elements (representing the fibre tows) connected to the nodes of the mesh discretizing the whole geometry, and the second is based on the consideration of a sink term in the continuity equation, reproducing the effect of the absorption of resin by the fibre tows. In this section, computational results obtained using these approaches are compared. In particular, reference data [115] have been used to represent the first method, while a simulation model including the sink term was solved, assuming the same processing conditions, using the ANSYS CFX software package.

From a theoretical point of view, the combination of the mass balance, including the sink term (Eq. 5.4), and the Darcy model (Eq. 5.5), described at the beginning of this chapter, yields

$$\nabla \cdot \left(\frac{K}{\mu} \cdot \nabla P \right) = S. \qquad [5.6]$$

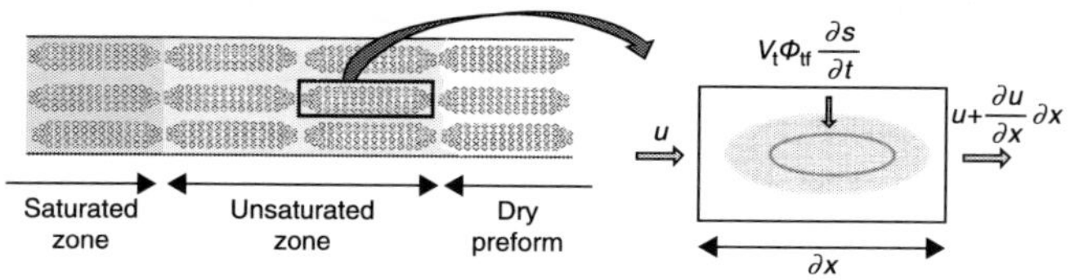

Figure 5.43 Mass balance in an elementary control volume including saturation effects

The sink term describes the resin flow from the inter-tow region to the intra-tow region, as depicted in Figure 5.43, assuming one tow in each control volume, and is a function of the saturation rate, since the saturation parameter s is defined as the ratio between the volume of resin inside the tow and the total volume that can be occupied by the resin (given by the product of the fibre volume fraction v_t, computed considering the tow as a solid, and the intra-tow porosity ϕ_{tf}).

Considering that the actual shape of the compressed tows is highly elliptical with a large aspect ratio, it is reasonable to assume them to be rectangular and to neglect the incoming flow from the sides. As a consequence, a one-dimensional flow through the single-scale fibre tow is derived (Figure 5.44), well described by the Darcy model (Eq. 5.5).

This assumption allows one to write the saturation s as the ratio between the location of the resin front inside the tow and half the tow thickness (for reasons of symmetry), i.e.

$$s = \frac{h}{h_t}, \qquad [5.71]$$

Simple manipulations yield

$$\frac{\partial s}{\partial t} = \frac{\partial h}{h_t \partial t} = \frac{K_t P_d}{h_t^2 \mu \phi_{tf} s}, \qquad [5.72]$$

where K_t is the tow permeability and P_d is the driving pressure (the liquid pressure outside the tow).

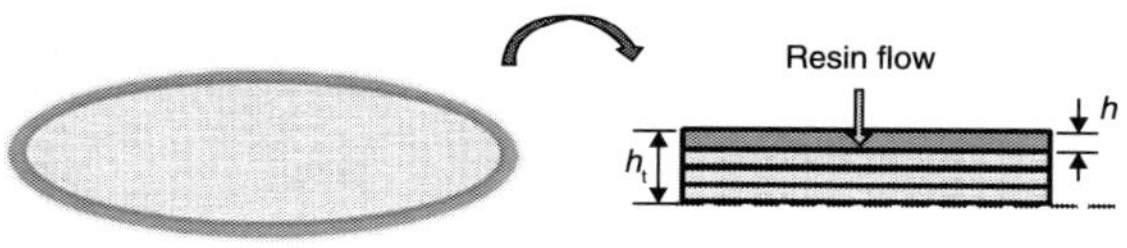

Figure 5.44 Tow saturation scheme

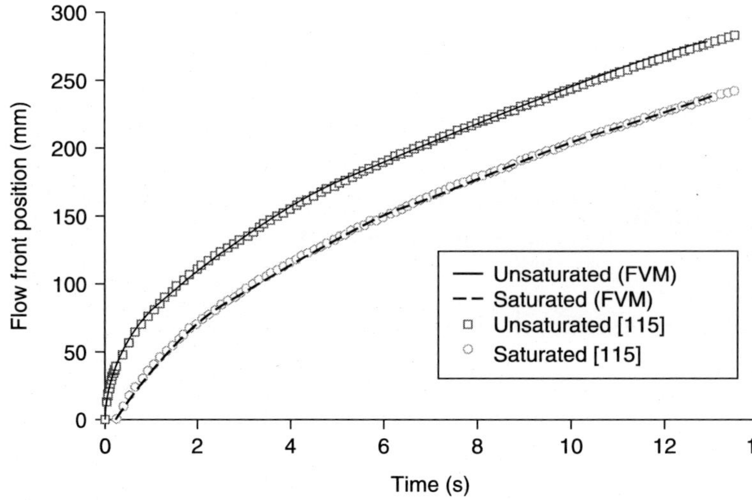

Figure 5.45 Numerical and analytically computed (using the mean viscosity value) resin flow fronts

In Figure 5.45, the positions of the unsaturated and saturated flow fronts, as a function of the impregnation time in the simulated one-directional flow test, are compared. The simulation parameters were as in Table 5.15, where P is the injection pressure, K_b is the bulk (macro-scale) permeability of the preform and L is the length of the preform. As can be seen, excellent agreement is found between the two models regarding the position of the unsaturated front, the beginning of saturation of the preform and the unsaturated flow length.

Table 5.15 Simulation parameters for the flow through a dual-scale preform

Parameter	Value
P (Pa)	10×10^5
μ (Pa s)	0.058
K_b (m^2)	1×10^{-9}
K_t (m^2)	1×10^{-12}
V_t	0.5
ϕ_{tf}	0.2
L (m)	0.288

5.2.6 Computational optimization of the pultrusion process

Introduction

In mathematics and computer science, an optimization problem consists in a search for the minimum (or maximum) point of a function $f: R^n \rightarrow R$ within an admissible set of solutions $\Sigma \subseteq R^n$. A detailed description of optimization procedures is beyond the aim of this work; however, it is relevant to outline the difference between derivative methods, based on evaluation of the gradient of the function considered, and derivative-free methods, based on an approximation of the derivative or only on the evaluation of function values [117, 118].

The latter class of methods, referred to as direct search methods, is widely applied to the optimization of engineering problems related to manufacturing processes. In these cases, the complexity involved often inhibits the obtainment of a closed-form solution connecting the process parameters to features of the final product or to the performance of the process. On the other hand, simplifying assumptions strongly reduce the effectiveness and applicability of analytical models. As a consequence, simulation-based optimization procedures are gaining considerable attention. In this section, the intriguing capabilities of these approaches are discussed, considering the case of the pultrusion process for the manufacture of thermoset matrix composite materials.

One of the most challenging tasks regarding the planning of a conventional pultrusion process is the definition of the best heating profile, i.e. the die temperature at each position along the pulling axis. The heating profile is imposed by proper regulation of the heating temperatures, i.e. the control temperatures of the electrical heaters; explicit reference to the heating system will be made in the applications considered in what follows. As a general rule, the die temperature should be sufficient to activate and control the cure reaction; it should provide a suitable distribution of the degree of cure in the cross-section of the manufactured piece and, at the same time, avoid thermal runaway and localized resin carbonization.

The influence of thermochemical aspects on the mechanical properties of the final product has been carefully analysed by Wilcox and Wright [119]. The same paper explores the use of some soft computing techniques, in particular artificial neural networks, for modelling the pultrusion process with the aim of process optimization, evidencing how the use of ANNs could lead to a greater understanding of the process, reduce product development times and increase the efficiency of the manufacturing process.

The relevance of appropriate planning of the heating profile, including also the pre-heater temperature, the heating-platen temperatures, the cooling temperature and the pull speed (which indirectly influences the steady-state temperature, owing to transport effects), has been highlighted in an experimental investigation performed by Carlsson and Åström [120], who remarked on the link between these features and the mechanical and surface properties of the processed part. What is more, some manufacturing defects, such as voids and micro-cracks, usually related to a non-optimal cure profile in the processed part, affect the linear and non-linear behaviour of pultruded composites [121]. Several computational models [62, 65–71, 74, 79, 84, 88] have been proposed to analyse the temperature and degree-of-cure profiles in a workpiece; however, very few studies have focused on the optimization of these profiles.

An optimization procedure for the pultrusion process based on the simulated annealing method was proposed by Coelho and Calado [122]. The objective function considered was formulated by taking into account an economic criterion and respecting some technological constraints related to the minimum value of the degree of cure and the peak temperature in the composite material. The solution was refined using a successive quadratic programming procedure. A fitness function was estimated for each potential solution by means of a thermochemical model of the process, solved using a Galerkin weighted residual finite element method [122]. The algorithm implemented proved to be capable of avoiding local minima. A different approach, based on the combination of a numerical model and mathematical procedures, was proposed in [72, 73]. Several process parameters, such as the heating-platen temperatures, the pull speed, the resin bath temperature and the cooling temperature, were considered in order to optimize the degree of cure of the material. Some limitations on the choice of the heating profile were taken properly into account by assuming that the processing temperature of the material could not exceed the degradation temperature of the resin system. This method was shown to be quite efficient; however, the output of the solution was affected by the combination of process parameters considered in the first step of the optimization procedure. In [123], a mixed integer genetic algorithm was employed to minimize the total number of heaters, satisfying constraints on the maximum temperature of the composite, the mean value of the degree of cure at the die exit and the pulling speed, in order to enhance the productivity of the process. An initial approach combining numerical models, ANNs and evolutionary optimization was discussed in [124]. In particular, computational results validated by considering a few test cases were used to train the network. Finally, an ANN was linked to a fast elitist non-dominated sorting genetic algorithm to optimize the die temperature and pull speed in pultrusion. In [125], a particle swarm optimization algorithm was coupled with a CFD model of the process, which was solved using a finite volume scheme to define the heating configuration in a pultrusion die, thereby reducing the energy requirements, evaluated in terms of heating rate. Regardless of the method adopted, however, the content of the papers cited here evidences the opportunities for the use of simulation-based optimization procedures, owing to the numerous potential solutions that need to be considered until algorithm convergence.

The optimization problem and algorithms

In this section, a hybrid method [126, 127] for the definition of a die heating profile that optimizes some features of a pultruded product is discussed. The procedure was based on the sequential use of genetic algorithms [128, 129] and the simplex method [130]. The objective (or fitness) function considered was based on the variance of the degree of cure, evaluated at the die exit cross-section, with respect to a target value. Analytically, this function can be written as

$$f = \sum_{i=1}^{N} \frac{(\alpha_{\mathrm{d}} - \alpha_i)^2}{N-1}, \qquad [5.73]$$

where α_{d} is the desired value of the degree of cure, *ai* is the degree of cure at the *i*th node in the final cross-section of the composite and *N* is the number of nodes in the cross-section. The practical meaning of this function is that it can be used to evaluate the combination of parameters required to obtain a uniform, satisfactory distribution of the degree of cure at the end of the process.

The post-die region was not included in the analysis, even though, as highlighted earlier, some enhancement of the degree of cure may be observed in the zone outside the die. However, the method described can easily be extended to include this effect. The objective function was evaluated and minimized by an iterative procedure based on a combination of the optimization algorithms mentioned above and the finite difference thermochemical model of the process described in Section 5.2.3. The choice of this numerical scheme was motivated by the stability of the calculation at high Péclet number [66, 74], which ensured a good trade-off between the computational time required for the evaluation of each potential solution and the reliability of the solution itself. The temperature and degree of cure profiles in the material being processed were numerically predicted using the finite difference model, and the fitness score of each candidate solution was then calculated using the degree of cure distribution in the final section.

It should be borne in mind, however, that in pultrusion the control parameters that regulate the die temperature profile are the temperatures of the heating platens. In several cases, the control system allows electric currents to flow in resistors (dissipating heat by the Joule effect) depending on the temperature measured by thermocouples placed on the heaters, rather than in the proximity of the workpiece. As a consequence, heat transfer phenomena inside the heating die must be properly considered in the optimization problem by including also the die in the computational domain, and therefore solving a discretized form of Eqs 5.46, 5.47 and 5.50.

A suitable penalty factor p_f was included in the fitness function to account for possible degradation of the resin system if the carbonization temperature is experienced by the material. In other words, if the peak temperature in the material being processed exceeded the resin degradation temperature, the fitness score was incremented proportionally, reducing the interest in the potential solution considered. The modified objective function was therefore written

$$f = \left(\sum_{i=1}^{N} \frac{(\alpha_d - \alpha_i)^2}{N-1} \right) p_f. \qquad [5.74]$$

The finite difference model was coupled with a genetic algorithm. In general, genetic algorithms operate on a population of individuals, called chromosomes, which represent possible solutions to a specific problem. Each chromosome is characterized by a fitness score, which represents a mathematical description of the problem and measures the goodness of the solution considered. Individuals with a higher fitness score have a higher probability of reproducing, resulting in a new population of individuals characterized by the features of different parents. Individuals with a bad fitness score have a lower probability of reproducing, and therefore become extinct. The reproduction of any given population is based on the following fundamental operators: selection, crossover and mutation. Since crossover between the best individuals in the current population is favoured, the more interesting areas of the search space are analysed in looking for the solution. By repeating the above procedure, the good features of the parents are extended to the whole population after several generations.

The genetic algorithm that was implemented generated an initial population of individuals randomly, where each chromosome was defined by one possible combination of the temperatures of the heating platens. This meta-heuristic approach allowed investigating the entire search domain, avoiding local minima, in a limited number of generations. In this framework, the value of the penalty factor was very important, because a low value could not assure satisfaction of the condition imposed on the peak temperature, while an excessive value could reject near-optimal solutions. The robustness of genetic algorithms is due to the mixing of genes, which allows them to overcome local minima. It should be noted, however, that if the evolutionary process relies on a stochastic basis, it cannot be excluded that reproduction will lead to a new generation characterized by lower fitness scores. In general, this is not always considered as a negative event; however, some of the best individuals can be lost. In the present application, the best individual in the current generation was always preserved by direct copying in the successive generation.

When the solution had converged to a good enough (but not rigorously optimal) solution, the simplex method was applied [130], using the same fitness function and penalty factor. The need for this strategy can be well understood if we take into account that the simplex method is strongly influenced by the starting point and does not ensure that local minima are escaped from. However, starting from a near-optimal solution, it is faster than genetic algorithms.

The simplex algorithm [130] is one of the earliest and best-known optimization techniques. From a theoretical point of view, a simplex is defined as a convex geometrical entity in R^n characterized by n dimensions, with $(n+1)$ vertices $x_j \in R^n$ and a minimum number of edges, whose dimension is $(n-1)$, equal to $(n+1)$. Typical examples are the triangle in the two-dimensional space R^2 and the tetrahedron in the three-dimensional space R^3. According to the Spendley simplex algorithm, at each iteration, the worst vertex of the simplex is replaced by a new vertex, which is obtained by reflection in the case of an improvement of the solution, otherwise a reduction of the simplex with respect to its primary vertex happens. As a consequence, only variations in the dimension of the simplex are permitted, and its shape is always conserved. The basic idea of Spendley was later modified by Nelder and Mead [131], who introduced new possible operations, namely expansion and (inside or outside) contraction. According to the Nelder–Mead algorithm, shape and dimension variations are allowed during a walk, following these schematic steps:

- Sorting of the vertices of the simplex in descending order:

$$f_1 \leq f_2 \leq f_3 \leq \ldots \leq f_n \leq f_{n+1} \qquad [5.75]$$

where f_j denotes the value of the objective function at the vertex x_j.

- Reflection of the simplex; the worst vertex (x_{n+1}) of the current simplex is reflected with respect to the centroid x_c, and the reflected vertex x_r is evaluated as

$$x_r = x_c + p(x_c - x_{n+1}), \qquad [5.76]$$

where ρ is the reflection coefficient and

$$x_c - \frac{1}{n}\sum_{i=1}^{n} x_i. \qquad [5.77]$$

If $f(x_r) \leq f(x_1) < f(x_n)$, the reflected vertex is accepted, replacing the worst vertex, and the current iteration is terminated; otherwise, if $f(x_r) < f(x_1)$ or $f(x_r) \geq f(x_n)$, the simplex is subjected to an expansion or a contraction, respectively.

- Expansion of the simplex: the expansion point x_e is given by

$$x_e = x_c + \chi(x_r - x_c), \qquad [5.78]$$

where χ is the expansion coefficient. If $f(x_e) < f(x_r)$, the expanded vertex is accepted, otherwise the reflected vertex is accepted and the current iteration is terminated.

- Contraction of the simplex: if $f(x_r) \geq f(x_n)$, the simplex contracts. In particular, if $f(x_r) < f(x_{n+1})$, the outside contraction point x_{co} is evaluated as follows:

$$x_{co} = x_c + \gamma(x_r - x_c). \qquad [5.79]$$

Then, if $f(x_{co}) \leq f(x_r)$, the new vertex is accepted; otherwise, a reduction of the simplex is applied. If $f(x_r) \geq f(x_{n+1})$, the simplex contracts inside:

$$x_{ci} = x_c - \gamma(x_c - x_{n+1}), \qquad [5.80]$$

where γ is the contraction coefficient. Then, if $f(x_{ci}) < f(x_{n+1})$, the new vertex is accepted; otherwise, a reduction of the simplex is applied.

- Reduction of the simplex. The dimensions of the simplex are reduced, preserving only the best vertex, by the definition of n new vertices v_i, as follows:

$$v_i = x_1 + \sigma(x_i - x_1), \tag{5.81}$$

where σ is the reduction coefficient, and the procedure restarts from the vertex sorting. The coefficients ρ, χ, σ and γ were taken as 1, 2, ½ and ½, respectively.

The walk scheme described above was adopted here to refine the solution provided by the genetic routine until a predefined stopping criterion, based on the dimensions of the simplex, was satisfied. In other words, the algorithm stopped when the following equation was satisfied:

$$\frac{\left(\sum_{i=1}^{n}(f_i - \bar{f})^2\right)^{1/2}}{\sqrt{n-1}} \leq \varepsilon, \tag{5.82}$$

where ε was the tolerance imposed on the algorithm and

$$\bar{f} = \sum_{i=1}^{n} \frac{f_i}{n}. \tag{5.83}$$

More details and some limits on the ability of the algorithm to escape from local solutions, in particular for multi-dimensional problems, were discussed in [132–134]. The routine described was implemented in the MATLAB environment.

Numerical results

The hybrid procedure described above was applied to define the optimal combination of heater temperatures in a pultrusion process for a fibreglass/epoxy C-section workpiece, already used as a modelling benchmark in Section 5.2.3 and shown graphically in Figure 5.46. Results provided by the finite difference model and by the finite element model (respecting the restriction on the Péclet number) were compared for similar processing conditions. The optimization technique was applied using only the temperatures of the heating platens as optimization variables; all other parameters, such as the pull speed and the temperature of the cooling channel, were assumed to be constant in each test case. The geometry of the model and the spatial discretization of the cross-section for the finite difference and finite element models can be found in Section 5.2.3. Only half of the model was considered for symmetry reasons.

In the first step of the analysis, some genetic operators were studied, taking the resin bath temperature T_{bath}, the cooling temperature T_{cool}, the resin degradation temperature T_{rd} and the pull speed v_{pul} as 45°, 50°,

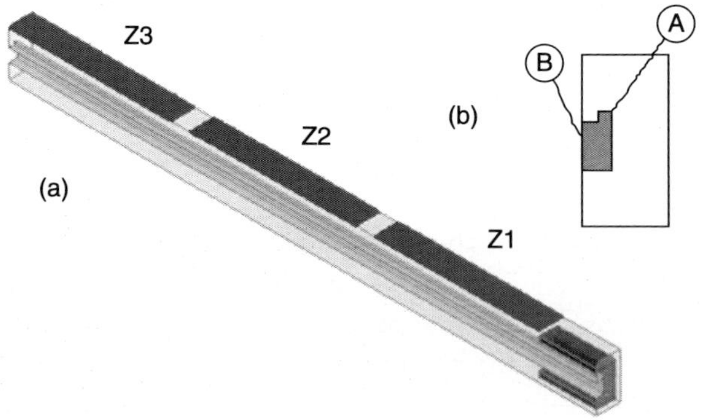

Figure 5.46 **(a) Pultrusion process and (b) section considered for optimization**

240 °C and 2.3 mm/s (resulting in a Péclet number equal to 38.7), respectively. The heating-platen temperatures were denoted by T_1, T_2 and T_3 for the heating platens on the top of the die, from die entrance to die exit, and T_4, T_5 and T_6 for the heating platens on the bottom of the die, in the same order. In this way, the heating profile was regulated to define the temperature in three zones (Z1, Z2 and Z3), moving from the inlet towards the outlet (see Figure 5.46). The range of the heating-platen temperatures, for the first generation, was chosen as [140, 220] °C. The lower limit was defined, for the specific resin system considered, by taking into account that the activation of the resin reaction is related to the amount of heat provided to the part by the platens. The upper limit was defined by considering the degradation temperature. It should be noted that the above range was not taken as a rigid constraint on the search domain; regions outside this range could also be explored. However, a suitable definition of the limits of the domain resulted in faster convergence of the algorithm. The desired value of the degree of cure in the objective function was 0.9.

Before running the optimization routine, a trial simulation was performed, assuming the temperatures of all the heating platens to be 180 °C, i.e. in the middle of the range considered. In Figures 5.47 and 5.48, the temperature and degree-of-cure profiles provided by the finite difference (FDM) and finite element (FEM) models of the process, as well as the distribution of the degree of cure in the cross-section of the composite material at the die outlet, are shown. The numerical results agree fairly well, evidencing the possibility of using the FDM as a first approximation for the optimization routine, even though it violates the restrictions dictated by the Péclet number. At point A, the temperature

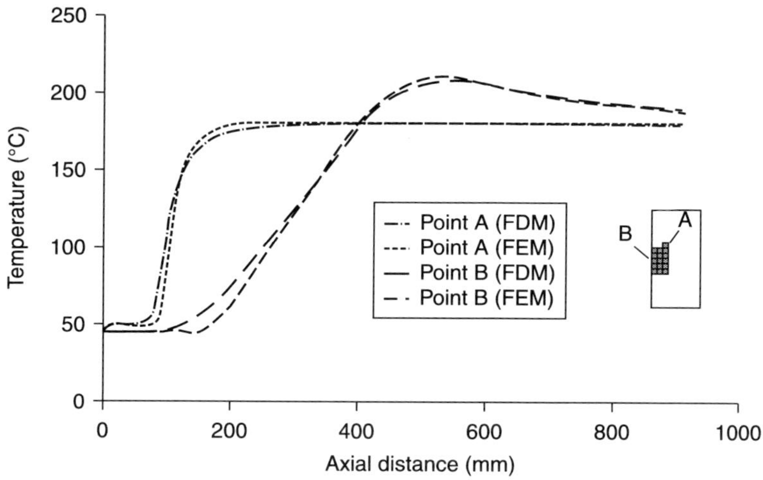

Figure 5.47 Temperature profiles in the pultrusion die in the reference case

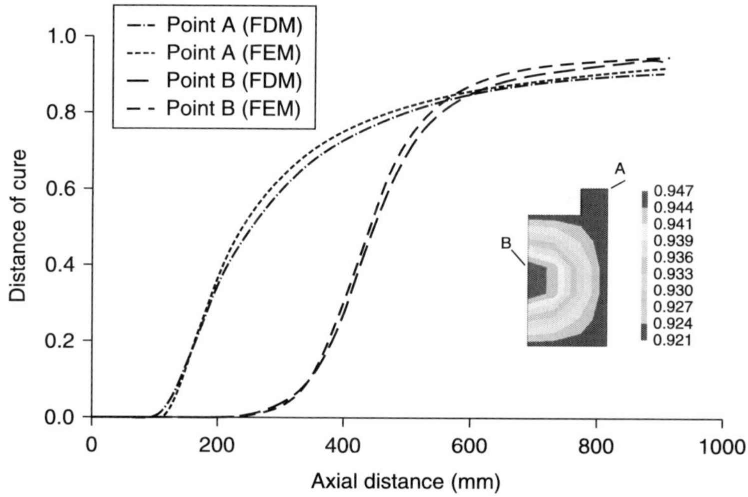

Figure 5.48 Cure profiles and degree of cure distribution in the final cross-section of the workpiece in the reference case

profile of the material resembles that of a sort of isothermal cure process, being practically imposed by the (equal) temperatures of the platens. At point B, progressive conductive heating and the effect of internal heat generation are more evident, leading to a temperature crossover at approximately 400 mm from the die entrance. The conversion profiles show an earlier activation of the cure reaction at the composite surface (point A); however, after the beginning of the cure reaction, at the core of the workpiece, the difference in the degree of cure is reduced until the cure crossover, at 590 mm from the inlet. What is more, the higher temperatures reached at B, which are attributable to the heat of reaction and exceed the die temperature, promote further cross-linking of the monomers. Excluding the post-die region, at the end of the simulated process the degree of cure was equal to 0.95 at B and 0.92 at A, both values being higher than the target value.

In Table 5.16, the mean value of the degree of cure α_m in the final cross-section, its standard deviation α_{std}, the value of the peak temperature T_p and the fitness score f_s are summarized for the reference case. As can be seen, the peak temperature predicted by the FEM is slightly higher than the value obtained using the FDM, but still lower than the critical value.

The first step of the optimization procedure was based on a standard genetic algorithm, assuming a population of 20 individuals for each generation and a maximum number of generations equal to 30. These choices were motivated by the need to compress the computational time required to define the initial simplex configuration. Following the random generation of the first individuals, the reproduction of the population started from the selection of parents, which were copied into a mating pool. In such a procedure, as a general rule, individuals characterized by a higher fitness score are copied more times than individuals characterized by a lower score. Several selection techniques can be used. The simplest one is uniform selection (US), which is useful for debugging and testing, but is not a very effective search strategy. In roulette wheel selection (RS), each chromosome is imagined to be placed on a roulette wheel and occupies an area proportional to the probability of being selected (depending on its fitness score). This technique preserves genes which could show their capacity in successive generations, since even individuals with a lower fitness score can reproduce, and convergence of the algorithm is fast. Tournament selection (TS) is based on competition between pairs of individuals in the current generation. The winner of each competition

Table 5.16 Numerical results for the reference case of the considered pultrusion process

Model	T_p (°C)	α_m	α_{std}	f_s
FDM	208.7	0.918	8.9×10^{-3}	4.0×10^{-4}
FEM	211.7	0.928	8.2×10^{-3}	9.3×10^{-4}

is the chromosome with the higher fitness score; this individual reproduces, while the loser can be chosen for other fights. Tournament selection, however, is characterized by slower convergence. In Table 5.17, a comparison of the results obtained using these strategies, limiting the procedure to genetic optimization but adopting different selection strategies, is listed. The selection methods were studied adopting uniform crossover between parents and allowing gene mutation at a rate of 0.01.

The best fitness score was achieved using roulette wheel selection, although the other methods also provided interesting results in terms of the mean value of the degree of cure. However, as shown by the convergence diagrams in Figure 5.49, which shows the best and mean values for each generation, no significant improvements in the best solution were obtained using uniform selection, while tournament selection quickly forced the population to collapse onto the best chromosome in the first generation.

After selection, the genes of the chosen individuals were mixed; this operation is known as crossover and is generally considered the most important genetic operation. Indeed, it influences directly the convergence of the algorithm; however, excessive use of crossover can cause premature convergence to a local minimum (or maximum). In the literature, three methods of crossover are widely used: single-point crossover (SPC), two-point crossover (TPC) and uniform crossover (UC). In single-point crossover, two chromosomes are randomly chosen and, after probabilistic considerations, their genes can be exchanged starting from a random point. Two-point (or multi-point) crossover likens chromosomes to

Table 5.17 Results of genetic algorithm using different selection methods

Selection strategy	Heating temperatures (°C) $T_1-T_2-T_3$ $T_4-T_5-T_6$	T_p (°C)	α_m	α_{std}	f_s
US	176.5–155.9–196.8 167.4–161.7–186.4	196.8	0.9024	5.5×10^{-3}	2.3×10^{-5}
RS	176.5–155.9–194.6 164.7–163.9–188.2	195.1	0.9044	3.4×10^{-3}	1.1×10^{-5}
TS	172.8–155.9–196.8 167.4–161.7–186.4	196.8	0.9027	4.4×10^{-3}	1.9×10^{-5}

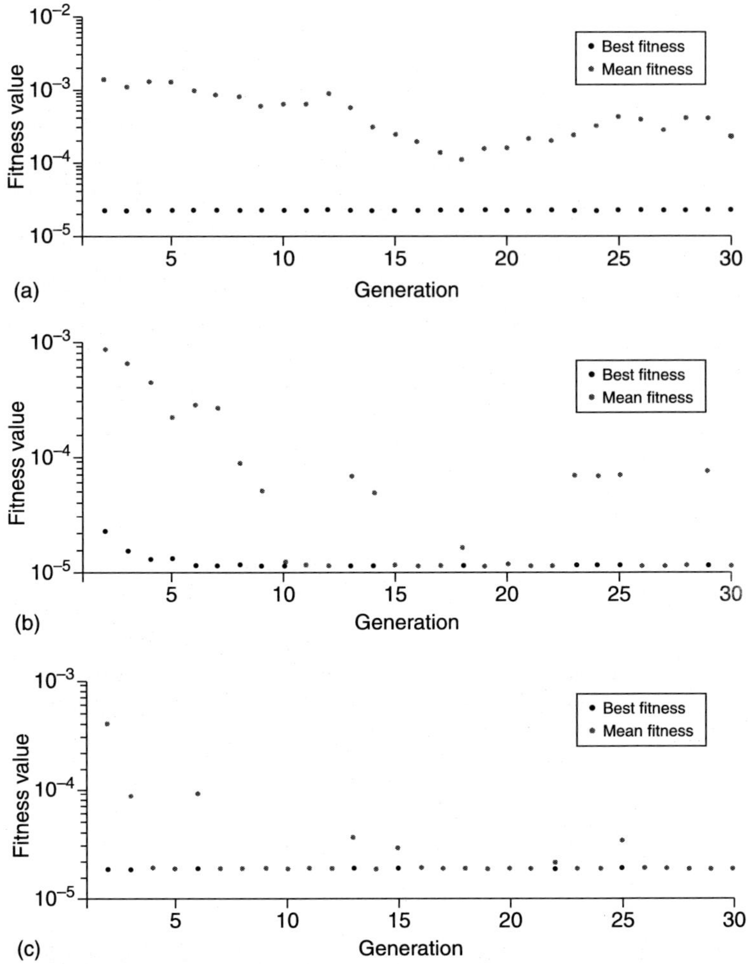

Figure 5.49 **Convergence plots using different selection criteria: (a) uniform, (b) roulette and (c) tournament**

circles; the procedure is similar to one-point crossover, but the two parents exchange only the genes between the two points. In uniform crossover, each parent gives a gene associated with a specific position to one of two children according to probabilistic considerations, and the other child takes the corresponding gene from the other parent.

In addition to selection and crossover, which define the 'natural' evolution of the population, the efficiency of exploration of the search space can be improved by allowing some genes of a new individual to mutate if some probabilistic conditions are satisfied. Mutation is a genetic operation that extends the search space towards zones not covered by the current generation. For each gene of a new individual, a random number is generated in [0, 1]; if this number is less than the probability of mutation, then the gene can mutate. Generally, if faster convergence can be achieved using crossover, mutation is a really useful operation when the search space is very large. Results obtained from a sensitivity analysis are listed in Tables 5.18 and 5.19.

Table 5.18 Influence of the crossover method on the results of the genetic algorithm (roulette selection)

Crossover strategy	Heating temperatures (°C) $T_1-T_2-T_3$ $T_4-T_5-T_6$	T_p (°C)	α_m	α_{std}	f_s
SPC	176.5–155.9–196.8 167.4–161.7–186.4	196.8	0.9040	4.8×10^{-3}	2.3×10^{-5}
TPC	172.5–155.9–196.8 182.7–160.2–185.4	196.8	0.8980	7.3×10^{-3}	2.1×10^{-5}
UC	176.5–155.9–194.6 164.7–163.9–188.2	195.1	0.9044	3.4×10^{-3}	1.1×10^{-5}

Table 5.19 Influence of the mutation rate on the results of the genetic algorithm (roulette selection and uniform crossover)

Mutation rate	Heating temperatures (°C) $T_1-T_2-T_3$ $T_4-T_5-T_6$	T_p (°C)	α_m	α_{std}	f_s
0.01	176.5–155.9–194.6 164.7–163.9–188.2	195.1	0.9044	3.4×10^{-3}	1.1×10^{-5}
0.05	165.0–155.9–196.8 181.3–161.7–186.4	196.8	0.8970	6.9×10^{-3}	1.2×10^{-5}
0.1	176.5–152.8–196.8 167.3–161.7–187.7	196.8	0.9030	4.2×10^{-3}	1.7×10^{-5}

The effectiveness of uniform crossover is evidenced by a better value of the fitness score, related to a more uniform distribution of the degree of cure. This can be explained by taking into account that this method involves complete mixing of genes between the two parents and that good individuals can be obtained easily. The optimal value of the mutation rate can be inferred to be 0.01; indeed, slow convergence and a higher standard deviation of the degree of cure were found using higher values of this rate.

It should be noted, however, that satisfactory (and similar) results were suggested by the genetic algorithm independently of the specific configuration adopted, proving the robustness of the proposed genetic optimization procedure. What is more, taking into account the subsequent refinement of the solution by the simplex method, it can be stated that for optimization of pultrusion processes, there is not an imperative need for careful setting of the operations and parameters of the algorithm.

A further improvement in the quality of the solution can be obtained by combining the above technique with the simplex method to refine the best individuals. At the end of the whole procedure, the fitness score was reduced by about one order of magnitude with respect to the pure genetic optimization procedure, and the degree of cure in the exit section was found to be very close to the desired value (Table 5.20).

Different temperatures were suggested for the heating platens placed on the top and bottom of the same zone of the die (Figure 5.50). This can be explained if we take account of the asymmetry of the cross-section of the part being processed, and the simultaneous contributions of heat transfer and heat generation to the promotion of the cure reaction. After the overall procedure, the higher temperature assigned to the bottom plate near the die entrance can be related to slow heat conduction into the composite material, while the lower temperature assigned to the

Table 5.20 Numerical results for the reference case of the pultrusion process

Method	T_p (°C)	α_m	α_{std}	f_s
Genetic algorithm (FDM)	195.1	0.9044	3.4×10^{-3}	1.1×10^{-5}
Genetic algorithm (FEM)	194.6	0.9075	4.5×10^{-3}	7.8×10^{-5}
Hybrid (FDM)	209.0	0.9015	1.0×10^{-3}	5.6×10^{-7}
Hybrid (FEM)	209.3	0.9023	1.3×10^{-3}	7.1×10^{-6}

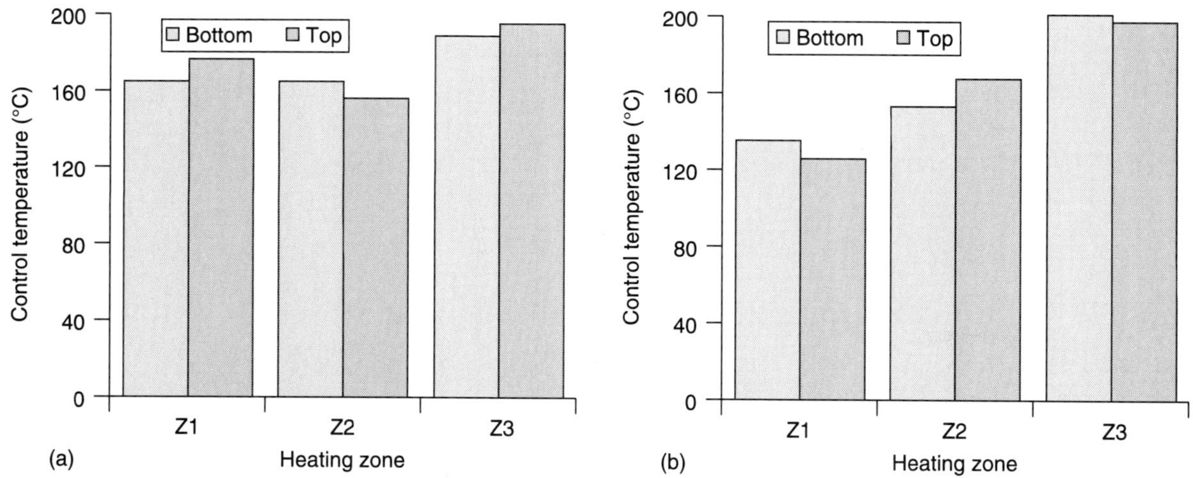

Figure 5.50 Control temperatures of the die heating zones, according to (a) the genetic optimization routine and (b) the hybrid routine

central plate at the bottom, with respect to the central plate at the top, can be explained by taking into account the heat generated by the resin cure reaction. In Figures 5.51 and 5.52, the numerical results provided by the genetic and full optimization procedures are depicted.

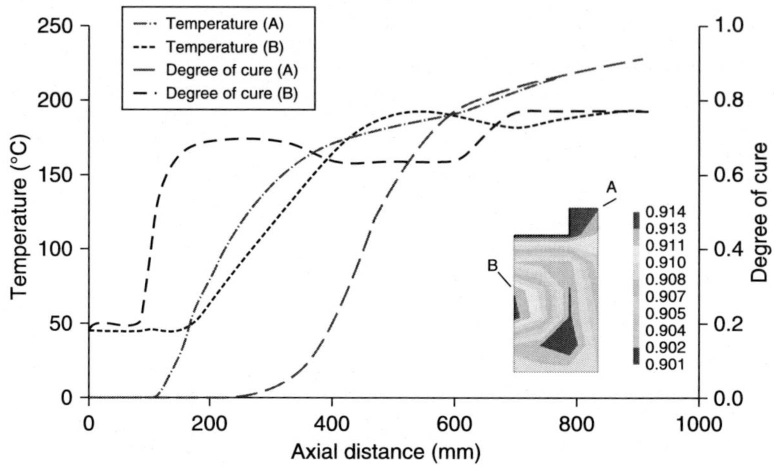

Figure 5.51 Temperature and cure profiles and distribution of the degree of cure in the final cross-section of the workpiece, after the genetic optimization routine using the FEM

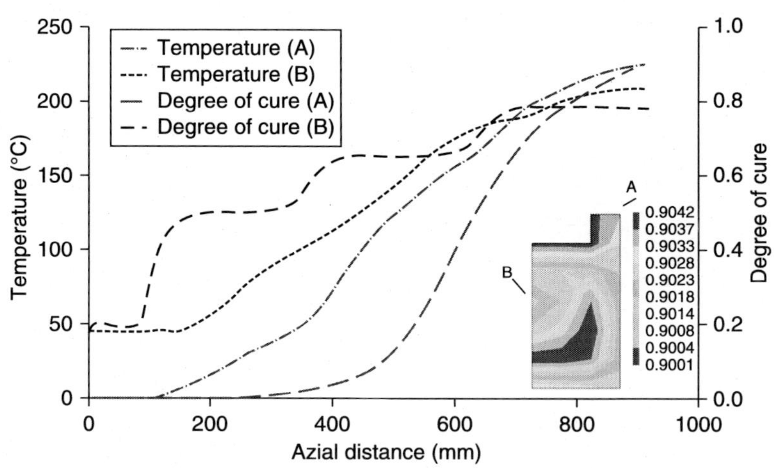

Figure 5.52 Temperature and cure profiles and distribution of the degree of cure in the final cross-section of the workpiece, after the hybrid optimization routine using the FEM

In the present case study, the optimization procedure was applied to obtain a part characterized by good quality, i.e. a uniform distribution of the degree of cure with a satisfactory mean value at the exit of the forming die. However, this procedure can potentially be applied, using appropriately defined fitness functions, to obtain different distributions of the degree of cure in the exit section, for instance in the case of post-die-shaping pultrusion. In this manufacturing process, the part being processed is completely formed outside the forming die and material cure is completed using ultraviolet radiation or other heating sources, and therefore the final extent of the reaction in the cross-section of the pultruded part is a crucial issue. The robustness of the method has been proved by the reliable results provided by the genetic algorithm, using all of the genetic operations. In particular, the optimal compromise between the convergence time and the fitness score was obtained using roulette wheel selection, uniform crossover and a rate of uniform mutation of 0.01.

Process optimization under different conditions

In this section, an application of the hybrid procedure described above for constrained optimization of the degree of cure in a pultrusion manufacturing process is reported. Different combinations of some process parameters were considered in order to investigate the capability of the method and to study the influence of the processing conditions on the optimal heating profile. In particular, a reduced optimization plan was defined, considering three factors (the resin bath temperature, the cooling-channel temperature and the pull speed) and two levels for each factor, as indicated in Table 5.21.

Table 5.21 Processing parameters for the optimized test cases

Test case	T_{bath} (°C)	T_{cool} (°C)	v_{pul} (mm/s)
Tc1	45	50	2.3
Tc2	40	50	2.3
Tc3	45	65	2.3
Tc4	45	50	7.5
Tc5	40	50	7.5
Tc6	45	65	7.5

Tables 5.22 and 5.23 summarize the mean values and standard deviations of the degree of cure in the final cross-section of the composite material, the peak temperatures, and the fitness scores obtained by simulating the optimal heating strategy suggested by the genetic algorithm and by the whole hybrid procedure. The numerical results were computed using the finite difference model, as applied in the computational routine.

As can be seen, the genetic optimization algorithm provides identical heating-platen temperatures in all cases characterized by the same pull speed, irrespective of the temperatures of the resin bath and of the cooling channel, highlighting the key role played by mass transport effects in the planning of the process. In Figure 5.53, the temperature and cure profiles at points A and B computed for the test case Tc4 are reported. Negligible differences were found relative to cases Tc5 and Tc6, so this plot can be considered as representative of all of the processes simulated with v_{pul} equal to 7.5 mm/s.

The heating profiles suggested by the genetic algorithm can be interpreted by taking account of the fact that, at higher pull speeds, the residence time of the material inside the curing die is significantly reduced; as a consequence, immediate activation of the resin reaction after the

Table 5.22 Heating-platen temperatures and results provided by genetic optimization using the FDM

Test case	Heating temperatures (°C) $T_1-T_2-T_3$ $T_4-T_5-T_6$	T_p (°C)	α_m	α_{std}	f_s
Tc1	176.5–155.9–194.6				
	164.7–163.9–188.2	195.1	0.9044	3.4×10^{-3}	3.1×10^{-5}
Tc2	176.5–155.9–196.8				
	167.4–161.7–188.2	196.8	0.9051	4.3×10^{-3}	4.5×10^{-5}
Tc3	176.5–155.9–194.6				
	164.7–163.9–186.4	194.6	0.9040	3.9×10^{-3}	3.1×10^{-5}
Tc4	216.0–214.5–195.8				
	207.1–219.1–219.0	219.1	0.7180	1.73×10^{-1}	6.33×10^{-2}
Tc5	216.0–214.5–195.8				
	198.2–219.1–219.0	219.1	0.6959	1.91×10^{-1}	7.85×10^{-2}
Tc6	216.0–214.5–187.5				
	207.1–219.1–219.0	219.1	0.7199	1.66×10^{-1}	6.02×10^{-2}

Table 5.23 **Heating-platen temperatures and results provided by hybrid optimization using the FDM**

Test case	Heating temperatures (°C) $T_1-T_2-T_3$ / $T_4-T_5-T_6$	T_p (°C)	α_m	α_{std}	f_s
Tc1	125.1–167.2–196.0				
	134.2–152.2–199.7	209.0	0.9015	1.0×10^{-3}	3.2×10^{-6}
Tc2	161.0–177.5–187.2				
	139.5–144.5–201.8	203.8	0.9018	1.7×10^{-3}	6.4×10^{-6}
Tc3	145.4–175.2–190.0				
	144.7–146.4–200.7	204.1	0.9017	1.3×10^{-3}	4.6×10^{-6}
Tc4	239.9–239.6–206.6				
	236.7–239.3–237.5	239.9	0.8533	9.3×10^{-3}	1.1×10^{-2}
Tc5	239.6–238.1–210.1				
	239.8–239.9–239.4	239.9	0.8455	1.0×10^{-1}	1.4×10^{-2}
Tc6	239.9–236.7–196.9				
	239.9–239.9–239.4	239.9	0.8551	8.8×10^{-2}	9.8×10^{-3}

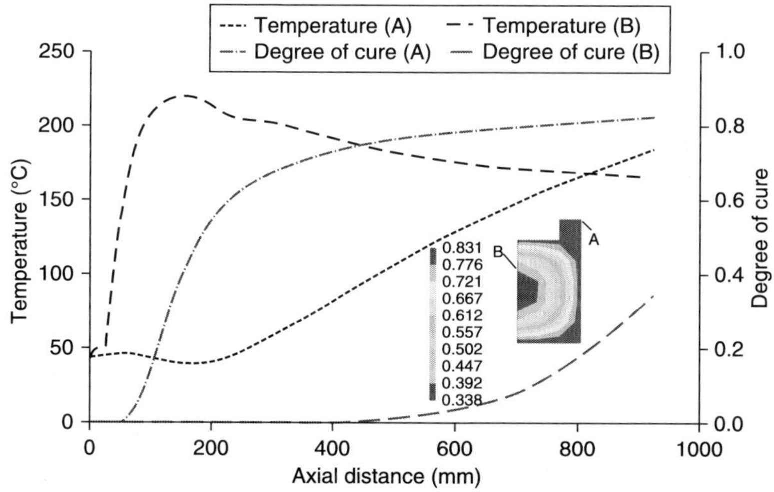

Figure 5.53 **Temperature and cure profiles and distribution of the degree of cure in the final cross-section of the workpiece after the genetic optimization routine (test case Tc4, using the FEM)**

inlet is needed, privileging the selection of individuals showing this feature. On the other hand, the similarity of the temperatures (obviously non-optimal in zones Z2 and Z3) highlights the limitations of the genetic algorithm with respect to refinement of the solutions, which reasonably requires numbers of individuals and generations considerably higher than those used in the test performed.

Comparing the two families of test cases (Tc1, Tc2, Tc3) and (Tc4, Tc5, Tc6), it can be observed that, for a pull speed of 2.3 mm/s, a progressive increase in the heating profile is suggested on moving from the die inlet to the exit. This heating strategy implies a progressive heat exchange from the die to the workpiece, which activates the resin reaction in an almost balanced way on the surface and at the core of the pulled material. Differently, when the workpiece is pulled at 7.5 mm/s, the shorter residence time of the workpiece in the die requires strong, fast heat exchange from the die to the workpiece. Lower temperatures need to be used for the other platens, according to the genetic algorithm, in order to avoid degradation of the resin by inversion of the heat flow from the workpiece to the die.

The intriguing capability of the simplex algorithm to improve the genetic solutions further can be well appreciated in Table 5.23, which clearly shows an enhancement of the mean value of the degree of cure in the cross-section of the composite at the die exit from 0.718 to 0.853, from 0.696 to 0.846 and from 0.720 to 0.855, in cases Tc4, Tc5 and Tc6, respectively. What is more, the significant slope of the cure profile and the temperature value at point B in the die exit (Figure 5.54), i.e. in the core of the workpiece, indicate that the material is still reacting, and therefore a further increase in the degree of cure should be expected. The peak temperature values increase from 219.1 to 239.9 °C in all of the aforementioned cases. In this regard, it is worth noting that, as computed by the finite difference model, the degradation temperature of the resin is never reached at any point in the composite material included in the simulation domain, underlining the robustness and the effectiveness of the procedure described.

The results point to the relevance of the pull speed to obtaining an optimized product, while a relatively minor influence is indicated for the resin bath and die cooler temperatures.

Analogous results computed using the finite element model with the same heating profile and processing conditions are provided for comparison in Tables 5.24 and 5.25. The need for the above validation was strictly related to the greater accuracy of the finite element model relative to the finite difference model.

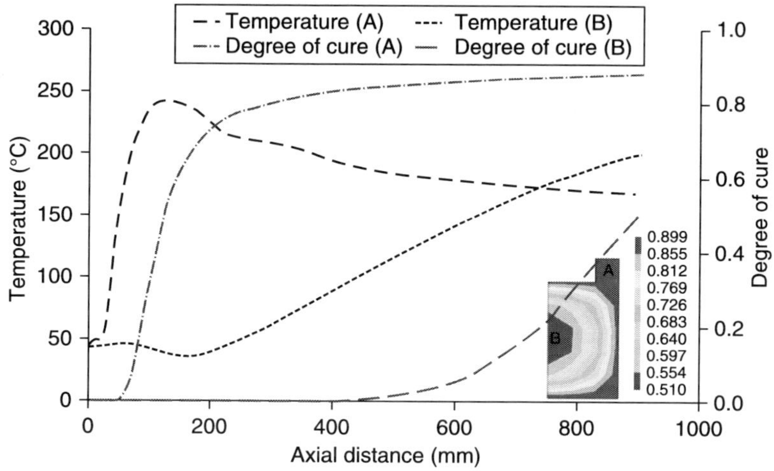

Figure 5.54 Temperature and cure profiles and distribution of the degree of cure in the final cross-section of the workpiece after the hybrid genetic routine (test case Tc4)

Table 5.24 Heating-platen temperatures and results provided by genetic optimization using the FEM

Test case	T_p (°C)	α_m	α_{std}	f_s
Tc1	194.4	0.9074	4.4×10^{-3}	7.8×10^{-5}
Tc2	195.5	0.9084	5.4×10^{-3}	1.0×10^{-4}
Tc3	195.1	0.9070	4.8×10^{-3}	7.6×10^{-5}
Tc4	217.0	0.7798	1.6×10^{-1}	4.1×10^{-2}
Tc5	217.0	0.7613	1.8×10^{-1}	5.3×10^{-2}
Tc6	217.1	0.7829	1.5×10^{-1}	3.8×10^{-2}

Table 5.25 Heating-platen temperatures and results provided by hybrid optimization using the FEM

Test case	T_p (°C)	α_m	α_{std}	f_s
Tc1	209.3	0.9023	1.2×10^{-3}	7.1×10^{-6}
Tc2	203.1	0.9030	2.9×10^{-3}	1.8×10^{-5}
Tc3	203.8	0.9025	2.1×10^{-3}	1.1×10^{-5}
Tc4	245.7	0.8924	7.9×10^{-2}	3.3×10^{-3}
Tc5	245.3	0.8862	8.9×10^{-2}	4.2×10^{-3}
Tc6	245.9	0.8929	7.6×10^{-2}	3.0×10^{-3}

As can be seen, good agreement is found between the results obtained using the FDM and FEM numerical schemes for a pull speed of 2.3 mm/s; however, at higher pull speeds, some relevant differences, attributable to an excessive increase in the Péclet number (126.4 at 7.5 mm/s), can be observed. This aspect is quite relevant, since, according to the FEM, the peak temperature exceeds the degradation temperature of the resin system modelled in cases Tc4, Tc5 and Tc6, owing to the excessively high temperatures of the heating platens and the heat dissipated by the reaction.

As the Péclet number increases, the heat generation rate due to the resin reaction is underestimated by the FDM, reducing the peak temperature and avoiding penalization of the fitness score. What is more, a global reduction in the temperature field in the resin lowers the local reaction rate and, as a consequence, leads to an underestimation of the mean value of the final degree of cure, as well as its standard deviation.

These considerations point to two important conclusions:

- The hybrid (genetic plus simplex algorithms) procedure implemented here is stable and efficient, since it forces the solution towards acceptable search zones (the peak temperature is in range, according to the FDM).
- The definition of a computational grid violating the condition imposed on the Péclet number, in order to reduce the computational cost, is a questionable strategy, since the reliability of the solution could be invalidated. On the other hand, a finer grid for the FDM implies a large increase in the computational time required for the analysis of each potential candidate for the solution. This limitation could be overcome by coupling the optimization technique with a meta-model (ANN-based) of the pultrusion process, trained using outcomes provided by FEM (or FDM or FVM) simulations of the process, respecting all of the accuracy conditions.

5.2.7 Computational optimization of an autoclave curing process

Introduction

Until recently, autoclave curing cycles were typically planned using pre-series trial and error procedures, based on heating profiles indicated by the resin suppliers. These approaches are, in most cases, expensive in terms of time and cost and do not ensure the obtainment of an optimal product or maximization of the performance of a specific process.

Nowadays, such procedures seem no longer to be acceptable if one is to be competitive in the market, well justifying the remarkable research effort that has been focused on numerical modelling and simulation of the cure process, as already stated in a previous section. Furthermore, computational procedures for pre-series process/product optimization are very appealing in the field of composites manufacturing, owing to the remarkable benefits achievable [56, 72, 73, 89, 135–140]. Taking into account that the optimization of manufacturing processes involves the consideration of multiple conflicting objectives and technical constraints to formulate a suitable objective function, numerous potential solutions should generally be examined until convergence of the algorithm. Experimental approaches based on multi-level optimization for the definition of pressures, temperatures and processing times to enhance the mechanical properties of the final product require extensive and expensive analysis, which significantly reduces the convenience of the method [141].

In the study described in this section, two different meta-heuristic optimization methods, based on the simulated annealing algorithm and on genetic algorithms, were implemented, tested and compared to minimize the curing time during autoclave processing, under specific technological restrictions. In particular, some variants of the original algorithms were introduced to escape from local minima, to improve the quality of the solutions and to accelerate convergence. The fitness score for each candidate solution was computed by means of the one-dimensional finite element thermochemical model of the process described in Section 5.2.2. The numerical model and the optimization algorithms were implemented as routines using the ANSYS Parametric Design Language [142].

The optimization problem

The objective function that was minimized was defined by the time t_{cure} needed to cure a thick composite laminate. The thermal cycle can be described in a time–temperature diagram by connected linear segments, as shown in Figure 5.55. Here, N denotes the total number of steps, the initial temperature is taken as room temperature, the initial time is taken as zero, and the decision variables consist of the final temperature T_i and the duration t_i of each step. This means that $2N$ independent variables need to be considered. An additive penalty p_a was used to account for technological constraints, yielding the following fitness function:

$$f = t_{cure} + p_a = \sum_{i=1}^{N} t_i + \sum_{k=1}^{K} \max(0, g_k)\lambda_k, \qquad [5.84]$$

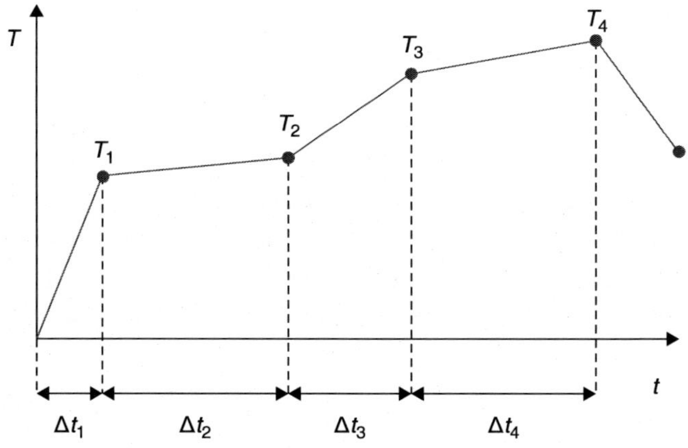

Figure 5.55 Representation of a four-step thermal cycle, including the final cooling

where K is the number of constraints considered, g_k is an integer-valued function that varies between 0 and K, and λ_k is a coefficient defining the penalty value to be added to the fitness score for each constraint violated.

During the cure process, a considerable amount of thermal energy is dissipated in the composite material because of the exothermic reaction. The relatively low thermal conductivity of the thermoset resins used prevents the generated heat from flowing away, inducing localized temperature increases and undesired thermal gradients. As a consequence, in hot spots, the temperature could exceed the carbonization temperature, leading to poor mechanical properties of the final product. Moreover, thermal gradients are responsible for thermal stresses and distortions of the part, which are generally more evident in the case of outside–in solidification after the gel point. Finally, it should be borne in mind that the mechanical properties of the processed part depend strongly on the final extent of the cure reaction and on the distribution of the degree of cure inside the part. The technological constraints adopted to define the optimal heating strategy were related to the transient temperature field, the direction of material solidification and the final distribution of the degree of cure in the manufactured product.

In particular, the value of the function g_k was increased by one unit if, during the process:

- the peak temperature in the material reached a critical value T_{max};
- the temperature difference between the surface and the core exceeded the allowed value ΔT_{max};

- the rate of temperature variation inside the material exceeded a critical value T_r;
- the degree of cure in the core was lower than that at the surface after resin gelation (outside–in solidification);

and, additionally, if at the end of the process:

- the mean degree of cure was lower than a minimum critical value $\alpha_{m,crit}$;
- the standard deviation of the distribution of the degree of cure exceeded a critical value $\alpha_{std,crit}$.

Two values of the coefficient λ_k were investigated to test the capability of the algorithm to detect penalty-free zones that would reduce the computational time, assuming $\lambda_1 = 10000$ s and $\lambda_2 = 20000$ s.

The properties of the material to be processed and the tooling, as well as the laminate thickness, were defined as in Section 5.2.2. An additional cooling step (by 50 °C with respect to the last temperature cycle within 10 minutes) was simulated for each potential solution, to stop the resin reaction and inhibit further heat generation and temperature increase. The penalty function was estimated assuming the following critical values: $T_{max} = 240$ °C, $\Delta T_{max} = 60$ °C, $T_r = 0.6$ °C/s, $\alpha_{m,crit} = 0.95$ and $\alpha_{std,crit} = 0.1$. Moreover, after resin gelation (at $\alpha_g = 0.6$), inside–out material solidification had to be induced.

Simulated annealing

The simulated annealing algorithm is an optimization method which mimics the slow cooling of metals, which is characterized by a progressive reduction in the atomic movements that reduce the density of lattice defects until a lowest-energy state is reached [143]. In a similar way, at each virtual annealing temperature, the simulated annealing algorithm generates a new potential solution (or neighbour of the current state) to the problem considered by altering the current state, according to a predefined criterion. The acceptance of the new state is then based on the satisfaction of the Metropolis criterion, and this procedure is iterated until convergence.

For this application, the Nelder–Mead algorithm, as described in Section 5.2.5, was employed to individuate new candidates for a solution, by probabilistically conditioning the acceptance of each movement of the simplex. In other words, during the annealing process, each new solution x_j was accepted with a temperature-dependent probability P_T given by

$$P_T = \begin{cases} 1 & \text{if } f(x_j) \le f(x_i), \\ e^{\left(\frac{f(x_i)-f(x_j)}{kT}\right)} & \text{if } f(x_j) \ge f(x_i), \end{cases} \qquad [5.85]$$

where T is the current temperature, k is the Boltzmann constant, and $f(x_i)$ and $f(x_j)$ are the fitness scores of the worst vertex xi and new vertex of the simplex, respectively. The annealing temperature varied within a range $[T_{in}, T_{fin}]$, following a predefined cooling rate c and a predefined cooling scheme.

At the beginning of the process, which started from a randomly generated simplex, the higher temperature and higher probability of acceptance of new solutions allowed the simplex to explore a wide region of the search space, thereby escaping from local minima; however, as the temperature was reduced, the probability of acceptance of unfavourable solutions was reduced. The simulated cooling process followed a geometric temperature reduction scheme, activated after a maximum number of simplex walks U_{max} or a maximum number of conditioned acceptances V_{max}. The algorithm parameters T_{in}, T_{fin}, c, U_{max} and V_{max} were defined as 2000 °C, 0 °C, 0.95, 20 and 5, respectively. Two stopping criteria, based on the maximum allowed number of simplex walks (250) and the minimum allowed difference between the best and worst fitness scores (100 s), were adopted. Four optimization routines were run for each configuration of the algorithm, to assess the repeatability of the method.

In Figure 5.56, one of the optimal thermal cycles evaluated by the algorithm that we implemented is depicted, showing also the temperature and degree of cure profiles experienced by the composite material. The first two steps of the cycle describe two sharp heating ramps from room temperature up to about 125 °C (at the same heating rate), which activate the cure reaction but avoid excessive thermal gradients and heating rates. Then a very slow increase in the autoclave temperature is indicated, in order to keep the peak temperature at an acceptable value. In addition, during the third step, an effective activation of the cure reaction in the core of the material is induced (after approximately 20 min), resulting in a cure crossover before the gel point ($\alpha = 0.45$, $t = 50$ min), with an inside–out direction of solidification.

Despite the validity of this result, the optimization campaign evidenced how the effectiveness of the simulated annealing algorithm in finding an interesting search space, as well as in escaping from local minima, was affected by the parameters characterizing the Metropolis criterion.

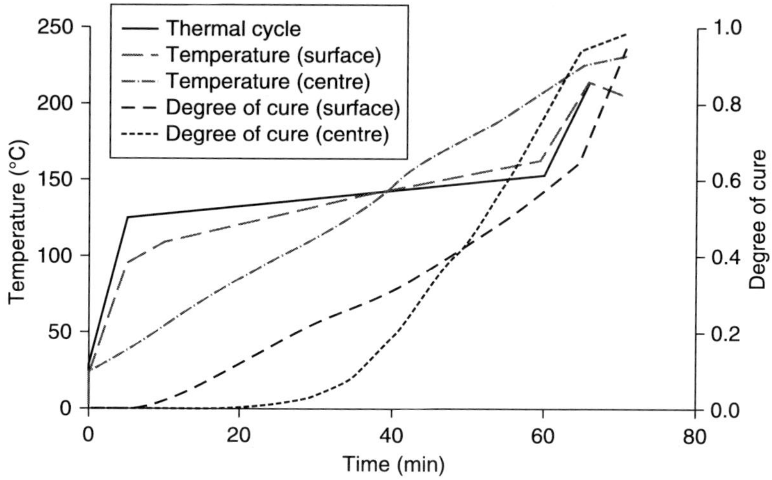

Figure 5.56 **Optimal thermal cycle and temperature and degree of cure profiles as suggested by the simulated annealing algorithm ($\lambda_k = \lambda_2$)**

Excessively high temperatures imply the acceptance of inappropriate movements, the repeated acceptance of reflected vertices and a simplex stall. On the other hand, low temperatures (or fast cooling) inhibit the capability of the simplex to escape, penalizing the convergence of the algorithm and the quality of the solution. Considering that the identification of the optimal algorithm parameters is quite complex and generally case-specific, a further modification of the combined algorithm was proposed to overcome any possible stall conditions. In this modified simulated annealing algorithm, the current worst vertex was replaced by a randomly generated new vertex if the Metropolis-conditioned acceptances led to a stall between two fixed simplex configurations for six consecutive walks.

In Figure 5.57, one of the thermal cycles suggested by the modified algorithm is shown. This appears quite similar to the previous suggested thermal cycle, highlighting the repeatability of the method using different variants. The main difference is observable at the beginning of the cycle, since two different heating ramps up to about 150 °C are exhibited. The subsequent plateau and the reduction in the external heat input allow the heat generated in the core to flow towards the surface of the material, avoiding resin carbonization.

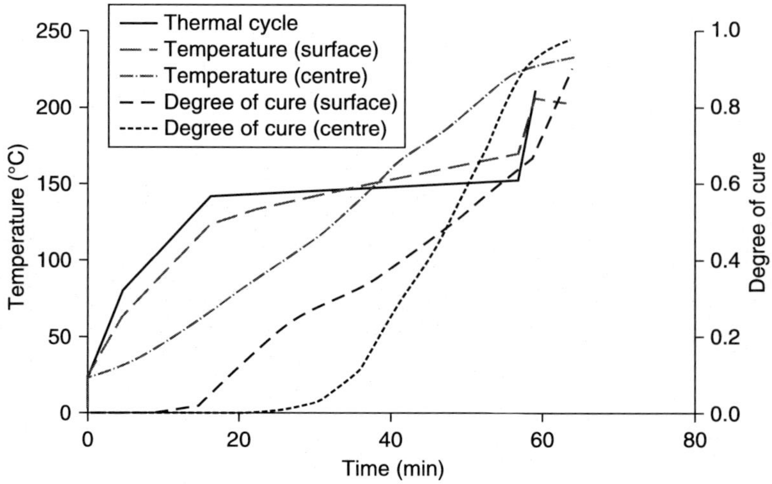

Figure 5.57 **Optimal thermal cycle and temperature and degree of cure profiles as suggested by the modified simulated annealing algorithm ($\lambda_k = \lambda_1$)**

An overall view of the performance of the two algorithms, in terms of fitness scores and computational times, is presented in Figure 5.58. As can be seen, the combination of the Nelder–Mead and simulated annealing methods is characterized by significant variability. Indeed, even if the Metropolis criterion allowed the simplex to escape from local minima in some cases, in other cases the simplex deviated towards worse search zones or was in a stall between two configurations. This behaviour can be explained by considering that, at the beginning of the routine, the simplex consists of several penalized vertices, whose reflection probably generates other penalized solutions, which are therefore characterized by reduced fitness score differences. As a consequence, the relatively high probability of acceptance at elevated temperatures forces the simplex to move between two stalling configurations. The subsequent cooling reduces the probability of unfavourable walks being accepted, thereby overcoming stall situations, but also limits the capability of the simplex to escape. The modification investigated allows the simplex to escape from local minima, and also to avoid deviation and repeated stalls into non-optimal regions. The value of the penalty coefficient appears not to be determinant.

Genetic algorithms

A general description of genetic algorithms has been given elsewhere in this text. For this application, two genetic algorithms, named GA1 and

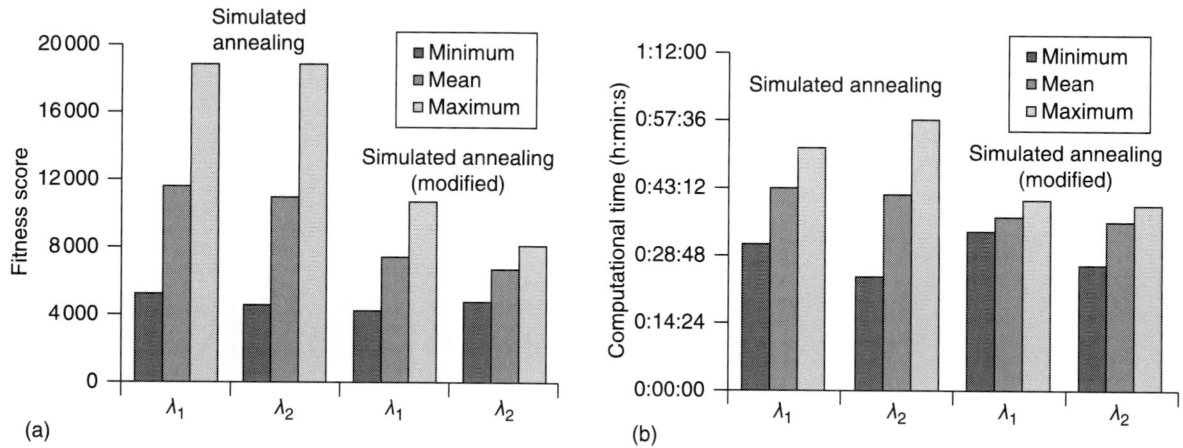

Figure 5.58 Performance of simulated annealing algorithm: (a) fitness scores at the end of the optimization and (b) computational times

GA2, were investigated. In particular, GA1 was based on the following scheme, where steps 2–6 were iterated until satisfaction of a stopping criterion based on the maximum number of generations (250):

Step 1. Random generation of the initial population (20 individuals).

Step 2. Evaluation of the fitness score of each individual.

Step 3. Roulette wheel selection.

Step 4. Two-point crossover.

Step 5. Mutation of a single gene according to a probabilistic criterion ($P_{mut} = 0.01$).

Step 6. Elitism by direct copying of the best individual into the successive population.

GA2 was derived from GA1, but included also two additional operators, referred to as substitution and swapping, specifically designed to take into account the contrasting effects of the technological constraints on the evolution of the algorithm in order to improve its performance. During a generic cure process, the cross-linking of monomers is favoured at relatively high temperatures; however, the constraint related to the peak temperature favours low thermal profiles. Furthermore, the requirement for inside–out solidification is easily satisfied by reducing the heating rate, but this increases the processing time. As a consequence, if only the selection, crossover and mutation operators were used, the identification of non-optimal but at least penalty-free solutions could require several generations and excessive computational time. Moreover, the basic version of the algorithm appeared quite sensitive to the value of the penalty coefficient: relatively low values attracted the population towards unacceptable zones, while high values led to the duplication of a current suboptimal solution and to entrapment in a local minimum.

The substitution operator, schematized in Figure 5.59, was applied with unity probability after a defined number of generations, to avoid excessive repetition of a single individual. Only two copies of the best individual were preserved for the successive generation, while removed chromosomes were replaced by randomly generated individuals. Premature convergence to a local minimum, caused by the repeated selection of substantially coincident individuals, could thus be avoided.

The swapping operator exchanged two randomly selected temperatures or durations describing the same cure cycle, after the satisfaction of a probabilistic criterion. The basic idea was to associate, in a single probabilistic operation, temperatures and durations related to different steps of the process to obtain a potentially interesting thermal cycle, as

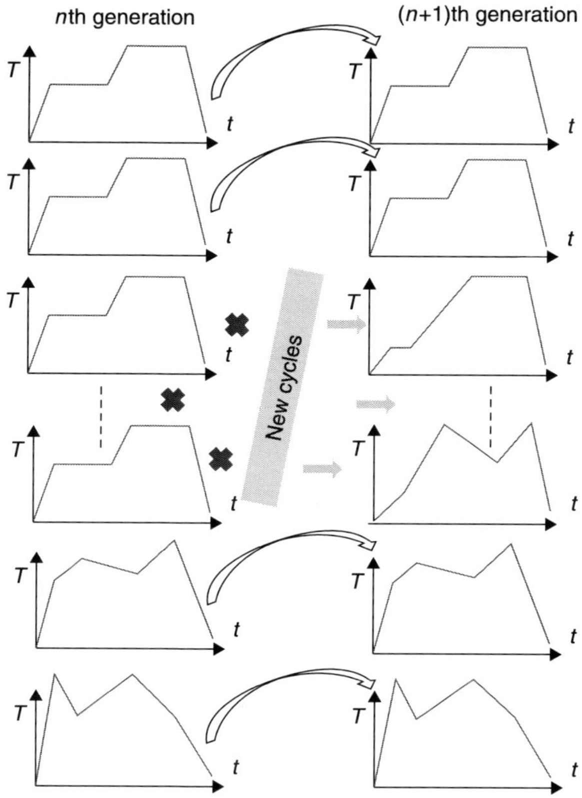

Figure 5.59 **Substitution**

shown in Figure 5.60. In this way, a remarkable compression of the computational time required to obtain similar results using only the standard operators was achievable.

In Figure 5.61, a cure cycle provided by the algorithm GA1 after 250 iterations is depicted, evidencing also the temperature and conversion profiles at different locations in the composite material. At first glance, the first steps of the optimized cycles are quite similar to the ones shown in Figures 5.56 and 5.57. A temperature increase from room temperature to 105 °C is indicated, to activate the cure reaction. However, differently from the other cycles, a sort of continuous heating (characterized by heating rates less than those at the beginning of the cycle) is suggested. This result is somewhat questionable: the external heat input during the third and fourth processing steps prevents the reaction heat from being easily dissipated towards the surface of the workpiece, which is already

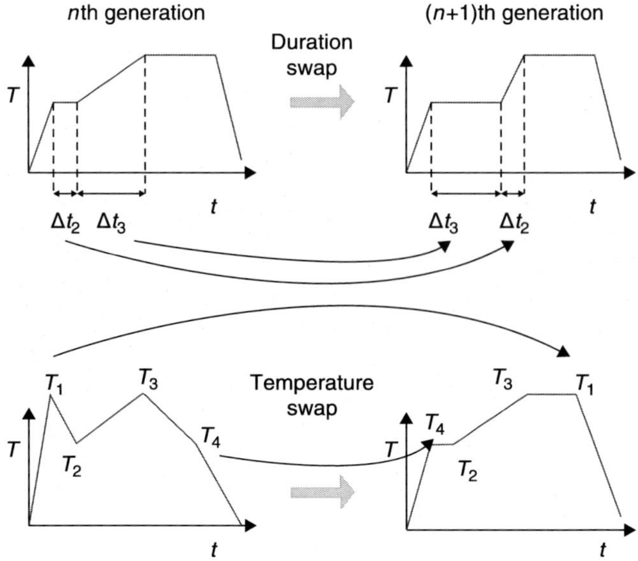

Figure 5.60 Swapping

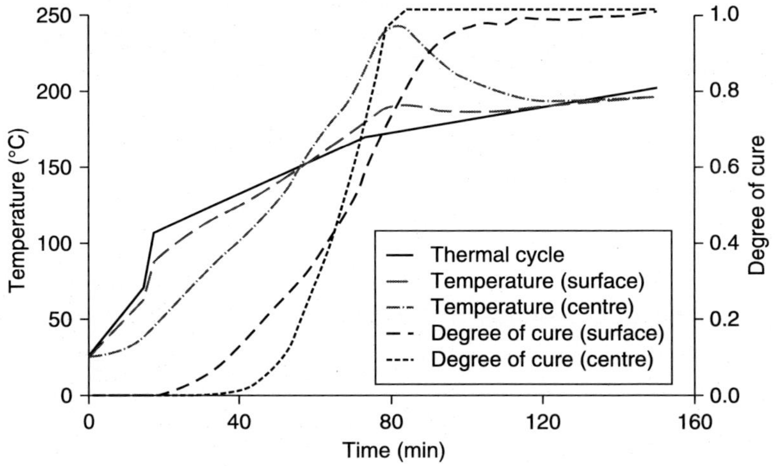

Figure 5.61 Optimal thermal cycle and temperature and degree of cure profiles as suggested by the algorithm GA1 ($\lambda_k = \lambda_2$)

heated by the external flux. As a consequence, the exothermic peak reaches values very close to (but still lower than) the critical temperature. To avoid resin degradation, the heating rate is controlled by increasing the durations of the two steps and, consequently, that of the whole cycle.

However, the algorithm does not incur any penalty and full cure is achieved throughout the thickness of the material at the end of the process.

The final result does not seem comparable with that of the routines previously described, since the cure time is approximately 150 min, versus 70 min in the best cycles obtained by applying simulated annealing and modified simulated annealing. What is more, since the number of individuals constituting each population exceeds the number of potential vertices of the simplex to be analysed during a walk, a significant increase in the computational time is observed (from 40 min to 9 h). The worsening of the optimization result can be related to the relatively slow evolution of the population if only the basic operators are activated. Indeed, taking into account the complexity of the problem considered, it should be expected that an improbable mutation of a single gene will provide an evident improvement in the solution. The mutated individuals will generally carry some penalty, requiring further refinement during successive generations to show advantageous features. This process is obviously slow, and the preservation of a mutated child (even if potentially interesting) is not guaranteed.

Optimization routines based on the algorithm GA2 were run following a strategy based on frequent substitution of duplicates (every 10 generations) and a relatively high swapping probability (0.3), in order to maximize the effectiveness of these operators. In Figure 5.62, one of the cycles provided is shown. The cycle depicted converges qualitatively and quantitatively to that shown in Figures 5.56 and 5.57, to a fair

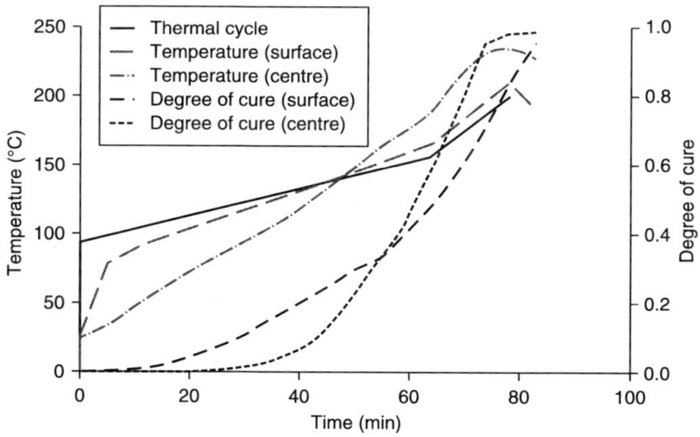

Figure 5.62 **Optimal thermal cycle and temperature and degree of cure profiles as suggested by the algorithm GA2 ($\lambda_k = \lambda_2$)**

approximation. Sharp heating ramps at the beginning of the process and to post-cure the material before the final cooling, spaced out by a longer step characterized by a low heating rate, are suggested again. This strategy implies good containment of the thermal and degree of cure gradients and a satisfactory position of the cure crossover before the gel point for $\alpha=0.33$ after 55 min from the beginning of the process. This clearly indicates the establishment of the desired inside–out solidification direction.

Figure 5.63 summarizes graphically the performance shown by the genetic algorithms that were implemented. As ascertained for the simulated annealing algorithm, the application of the genetic algorithm in its basic configuration (GA1) is affected by significant dispersion in the final results, with a slight worsening, increasing the value of the penalty coefficient. Moreover, the slow convergence and the longer cure cycle that was simulated resulted in higher computational costs. This is explainable by considering the relevance of the mutation operators to exploring new search spaces. Reasonably, higher penalty values strongly reduce the selection probability of mutated and penalized children, leading (sooner or later) to their extinction. As a consequence, the effectiveness of the operation is compromised. In contrast, for lower values of λ_k, mutated and penalized (but potentially interesting) individuals have a higher probability of being selected for crossover, possibly providing better children in successive generations.

The simultaneous activation of the substitution and swapping operations leads to an appreciable global improvement in the performance of the genetic algorithm, in terms of the quality of the solutions and the computational time required. What is more, it also reduces the sensitivity of the algorithm to the settings of some parameters, such as the mutation rate and the penalty value. The replacement of duplicate individuals allows the insertion of new genes into the population; possible advantageous features make these individuals suitable for selection, transmitting their heritage to successive generations. In this way, the population can quickly evolve.

Comparing Figures 5.58 and 5.63, it can be concluded that all of the methods described are able to identify penalty-free cycles, proving the robustness of the algorithms implemented. The algorithm based on simulated annealing, employing the simplex method for the definition of neighbours, is effective for escaping from local minima if a stall condition is avoided. The main advantage of this technique is related to the appealing compression of the computational time (about 36 min against the 6 h required by the genetic optimization routines on average). On the

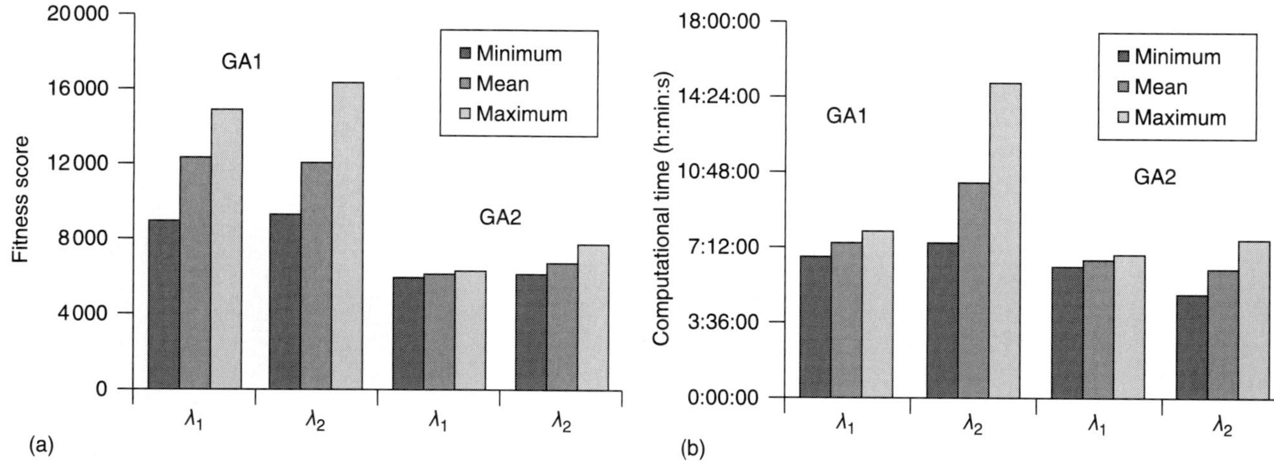

Figure 5.63 Performance of genetic algorithms: (a) fitness scores at the end of the optimization and (b) computational times

other hand, the great repeatability of the results computed by the genetic algorithm including the substitution and swapping operators is evident.

5.3 Soft computing and brake friction materials

5.3.1 Modelling and prediction of the cold performance of brake friction materials

Introduction

The braking performance of a vehicle is decisively influenced by the friction system used, which consists of a brake pad pressing against a cast iron disc. This affects safety, braking comfort (noise, vehicle vibration, pedal feel, smoke generation etc.) and service life. The demands imposed on the friction system, and particularly on the behaviour of brake pads, under a wide range of operating conditions are high and manifold. The capabilities of braking systems have to be constantly improved owing to different vehicle weights, the fact that some vehicles have four-wheel drive, different maximum speeds etc. This is additionally complicated by the introduction of add-on systems such as ABS, ASR, ESP and brake assist. It should be obvious that different vehicles impose different requirements on the brake friction material used as part of the braking system. One or two friction formulations are not sufficient to meet the aforementioned requirements for different families of vehicles. Therefore, it is necessary to develop friction materials that are able to satisfy all of the predetermined requirements for specific vehicle families. These requirements are especially related to (i) the need for a short bedding period, (ii) stability of the cold performance of the brake, (iii) stability of brake performance during speed changes, (iv) stability of brake performance under temperature load, (v) stability of brake performance after temperature load, (vi) stability of brake performance under changes in the activation pressure, (vii) the relation between the 'static' and dynamic friction coefficients, (viii) the friction coefficient under wet conditions and (viii) mechanical characteristics (compressibility, shear strength, bending strength etc.).

A wide range of ingredients are used by friction material manufacturers. According to [144], more than 2000 different raw materials and variants of them are now used. However, friction material manufacturers currently use only approximately 150 different ingredients to produce friction materials for braking systems. The types and relative amounts of ingredients in a commercial brake friction material are determined by considering many performance-related issues, such as the friction force, the propensity of the material to generate noise, its aggressiveness against rotors, brake-induced vibration and wear [145]. An individual automotive brake friction material usually contains 10–25 different raw material ingredients [146, 147] to meet the requirements for reliable and comfortable brake performance over a wide range of brake actuation pressure, temperature, humidity and sliding speed [146]. Since the number of requirements imposed on friction materials and the number of ingredients used in their production are continuously increasing, improved capabilities for the prediction of the characteristics of brake friction materials are highly desired.

The prediction of the behaviour of brake friction materials under different operating conditions is complicated by the fact that the braking process has a stochastic nature. The process is affected by changes in the size of the real contact area; the existence of a transfer layer between the friction pair; changes in pressure, temperature and speed; deformation; and wear. The performance of a brake system depends on the interaction between the brake rotor and brake lining at the sliding interface, involving complicated mechanical and chemical processes [148]. The size of the area of real contact between the pad and the disc is far from constant [147], it is very small compared with the total contact area [149], and it is highly dependent on changes in pressure, temperature, deformation and wear. The microscopic contact points change dynamically from place to place in fractions of a second during braking [149]. Owing to these very complex contact conditions, the effects of material properties on the sliding contact during braking need to be investigated [150, 151].

Besides the complex contact situation, it was established in [152, 153] that the formation and stability of transfer films on the counter-face play an important role in the friction and wear behaviour of polymers and polymer composites sliding against a metal surface. Furthermore, according to [154], the durability of the friction film at the interface at temperatures above the decomposition temperature of the binder resin appears to be very important for the brake performance and wear resistance of a friction material. The composition of the friction film influences mainly the frictional characteristics, but is not clear, according to [154], why transfer films are selectively formed. The thickness and surface morphology of transfer films are highly dependent on the temperature, the brake actuation pressure and the chemical state of the ingredients in the brake lining. This occurs because the cohesion of the constituents of the film and the rheology of the film are different under different sliding conditions [155]. This is important because the temperature at the friction interface strongly affects the properties of the transfer film, resulting in continuous variation of its thickness and composition as a function of the sliding time. According to [156], the frictional performance is independent of the thickness of the transfer film but sensitive to the composition of the film. It is obvious that the mechanism of formation of the friction film is very complicated and depends strongly on the thermal history of the sliding interface. Organic constituents, fibrous materials and solid lubricants play important roles in establishing a transfer layer at the friction interface [157].

It is evident, according to [146, 147, 152, 158], for example, that friction and wear processes are highly dependent on the ingredients of the friction material, and as a result of special requirements, as was pointed out in [159], friction materials have evolved into very complex structures. However, consistent scientific approaches to obtaining an optimum formulation for enhanced brake performance are difficult to find in the literature. As explained in [160], this is partly due to the difficulty of handling large numbers of experiments to obtain reliable frictional properties as a function of the amount of each ingredient. Hence, performance assessment and optimization is a decision-making task with conflicts. That is why attention was paid in [161] to the use of multiple-criterion optimization by a ranking and balancing method for the selection and design of an optimal composite. On the other hand, some optimizations of the composition of friction materials have been performed by trial and error methods, as suggested in [162]. According to [163], the two most important relationships are not yet known: the first of these is between the components of the mixture and the friction layer, and the second is between the friction layer and the frictional behaviour of the system.

Thus, the basic problem is how to develop an advanced, engineered friction formulation and how to define the most appropriate manufacturing parameters in a reduced time and at reduced cost, with an increased number of requirements related to the final performance of the friction material. The achievement of a satisfactory solution to this problem needs a modelling effort, in particular for prediction of the effects of changes in the formulation, manufacturing process and/or operation conditions of the friction material. It is evident that a method for simultaneous prediction of the influence of the friction material formulation and of the manufacturing and operation conditions is needed. According to [164, 165], the modelling of material properties generally involves the development of a mathematical model derived from experimental data. In this context, some soft computing techniques, such as artificial neural networks (ANNs), appear highly promising in the field of materials science. For instance, in [166] ANNs were applied as a tool for the optimum design of composite materials. The ANN technique has been applied to several different engineering problems, for example modelling of the wear of polymer composites [164], analysis of the dynamic mechanical properties of PTFE [165] and prediction of erosive wear of polymers [167].

The main deficiency of the investigations in the field of brake friction material engineering is that not all of the influencing factors (formulation, manufacturing and operation conditions) have been included in the prediction of the behaviour of brake friction materials. That is why an attempt has been made here to explore ANNs as a tool for modelling and prediction of the cold performance of brake friction materials. It is important to underline that these predictions were made by trying to integrate simultaneously the influences of the complete friction material formulation, the most important manufacturing parameters and the operation conditions of the friction material. The modelling and predictions were related to the cold performance of the friction material before and after fading and recovery tests.

Experimental data

As has been explained, very complex and highly non-linear phenomena are involved in the field of tribology. This is the reason why analytical models are difficult or even impossible to obtain. Artificial neural networks, described as 'computational models' [168], have been introduced to learn from and generalize experimental data based on parallel processing. The process of developing an artificial neural model is not trivial and involves many critical issues: (i) selecting a data generator, (ii) planning data generation (defining the range and distribution of the training data in the model input parameter space in which training data needs to be generated), (iii) performing data generation, (iv) data pre-processing, (v) selection of the neural-network structure, (vi) selection of the training algorithm, (vii) neural-network training, (viii) evaluation of the accuracy of validation and (ix) testing of the ANN [169].

At the beginning, a neural model of the behaviour of a friction material does not know any information about the performance of the material in different operating regimes. In order to learn about the behaviour of friction materials with different formulations and different manufacturing and testing (or operation) conditions, the ANN has to be trained with corresponding data. Therefore, a preliminary step in the development of a neural model is the identification of the model's input and output. This identification depends on the objectives of the model and the choice of data generator. According to the objectives considered in this chapter, the input parameters are defined by the formulation of the friction material, the manufacturing conditions used and the testing conditions (Figure 5.64), against changes of the friction coefficient or the brake factor C (a one-dimensional output parameter). The type of data generator depends on the application considered, the availability of data and the accuracy.

The brake performance resulting from the complex interrelated tribological processes occurring in the contact region of the friction pair during braking is affected by the physicochemical properties of the ingredients of the friction material. In order to establish a relationship between the input and output parameter spaces, the role of the data generator is important from the point of view of repeatability of the testing conditions. That is why it was decided to perform the testing of the friction materials under strictly controlled conditions, related to changes in the brake actuation pressure, the initial speed, the initial temperature and the inertia of the revolving masses. These testing

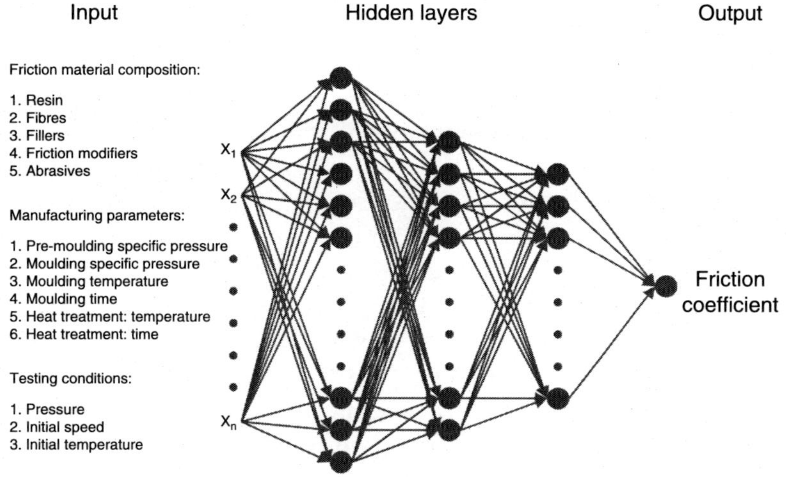

Figure 5.64 **Artificial neural network model for prediction of the behaviour of friction materials [170]**

conditions were chosen in order to simulate real operating regimes with a full-scale inertia dynamometer. The dynamometer was equipped with a PC-based automatic control and data acquisition system that measured the pressure, speed, temperature and braking torque at a sampling rate of 50 Hz. The brake factor C was calculated from average values of the friction coefficient determined as the speed changed in the range between $0.8v$ and $0.1v$, where v is the initial speed.

Therefore, using the inertia dynamometer as a data generator, the friction materials could be tested according to the testing methodology adopted. A testing methodology needs to be chosen according to the range and distribution of the data that are going to be collected. As shown in Table 5.26, there were five different tests. Our attention was paid to cold performance tests before and after a thermal load was applied to the brake. The testing methodology, according to Table 5.26, was divided into the following tests: (i) a burnishing procedure, (ii) 'cold performance 1' (identification of the synergistic influence of the brake pressure and the initial speed on the friction coefficient, with a cold brake – the initial brake interface temperature was less than 100 °C),

Table 5.26 Cold performance – testing methodology [170]

Test No.	Test	Pressure (bar)	Initial speed (km/h)	Temperature (°C)	Number of braking cycles
1	Burnishing	40	90	<100	100
2	Cold performance 1	20, 40, 60, 80, 100	20, 40, 60, 80, 100	<100	25
3	Fading	Corresponding to 3 m/s^2 deceleration in first braking cycle	90	Open	15
4	Recovery	20, 40, 60, 80, 100	20, 40, 60, 80, 100	Reached after 3 min cooling at 300°/min after fading test	25
5	Cold performance 2	20, 40, 60, 80, 100	20, 40, 60, 80, 100	<100	25

(iii) a fading test (identification of the influence of the brake interface temperature on the friction coefficient after 15 braking cycles with 45 s pauses between successive braking cycles), (iv) a recovery test (identification of the capabilities of the friction material to preserve its performance under the high-temperature load that it was subjected to after the fading test and 3 min of cooling at 300 °C/min) and (v) 'cold performance 2' (identification of the synergistic influence of the brake pressure and the initial speed on the friction coefficient after high-temperature load). This methodology offered intriguing possibilities for checking several characteristics of the brake friction material, namely the sensitivity of the friction coefficient to (i) pressure changes, (ii) changes in the initial speed, (iii) temperature changes simultaneously with pressure and speed changes, and (iv) pressure and speed changes after high-temperature load.

The importance of the ranges used for the input/output values and their distribution has already been mentioned. Once the range of the input parameters has been decided, the next step is to choose a sampling strategy. The range and distribution of the testing conditions are defined by the choice of testing methodology. However, the choice of the range and distribution of the manufacturing parameters of a friction material, especially its formulation, is a much more difficult task. The problem related to the range and distribution of the formulation parameters is complicated further by the fact that more than 150 different raw materials are used in the friction material industry today. That is why decisions about the selection of ingredients, the range of the proportions of those ingredients and their distribution have to be made taking into consideration the training, validation and test data sets. For the purpose of creating the training and validation data sets, the ingredients in the friction material formulation and the ranges of the proportions of each ingredient were selected as shown in Table 5.27 (the friction materials denoted by A, B, C and D). Owing to the unknown interrelated influences of these ingredients during braking, a random distribution of the proportions of the ingredients over a selected range was used. In order to evaluate the generalization capabilities of the selected artificial neural networks, the test data set was sampled by testing a new type of friction material. The formulation of this friction material (Table 5.27, friction material E) differed from those used for the training data set. A similar strategy was used for selecting the range and distribution of the manufacturing parameters, as shown in Table 5.28. It is evident from Table 5.28 that the manufacturing parameters used for friction material E were completely different from those used for materials A, B, C and D.

Table 5.27 Friction materials A, B, C, D and E – ranges of proportions of ingredients used (vol%) [170]

Ingredient	Friction materials A, B, C, D (training data set)	Friction material E (test data set)
Phenolic resin	17–20	
Iron oxide	3–5	2
Baryte	26–15	12
Calcium carbonate	1–3	
Brass chips	1–2	3
Aramid	4–6	7
Mineral fibre	12–16	10
Vermiculite	4–7	
Steel fibre	4–2	1.5
Glass fibre	2–3	4
Brass powder	1–2	0.8
Copper powder	1–2	
Graphite	7–4	
Friction dust	5–3	6
Molybdenum disulphide	1–3	
Aluminium oxide	2–3	3.5
Silica	1–2	
Magnesium oxide	8–2	

Table 5.28 Manufacturing parameters [170]

Manufacturing parameter	Friction materials A, B, C, D (training data set)	Friction material E (test data set)
Moulding pressure (MPa)	45–65	40
Moulding temperature (°C)	155–170	170
Moulding time (min)	6–11	10
Heat treatment temperature (°C)	200–250	200
Heat treatment time (h)	12–5	8

The five different types of friction material were manufactured and tested in a disc pad assembly. The disc pads were designed to be mounted on the front axle of a passenger car (static load 730 kg, effective disc radius 101 mm, floating caliper, piston diameter 48 mm), and had a friction surface of area 32.4 cm^2 and a thickness of 16.8 mm. The formulation and manufacturing parameters are shown in Tables 5.27 and 5.28. Each type of friction material was completely different from the others regarding the formulation and manufacturing conditions, while the testing methodology was the same (Table 5.26). The results obtained from the testing of the friction materials A, B, C and D were used for training the neural networks, while the results of the testing of material E were used for testing the networks. As shown in Table 5.27, the ranges of the proportions of the ingredients were not the same for the training and test data sets. The proportions (volume percentage) of all ten ingredients of friction material E took values outside the ranges used for the training data set (Table 5.27). This was done in order to test the generalization capabilities of the ANNs regarding unknown input data.

The disc pads were manufactured by a conventional procedure, which consisted of dry mixing, preforming, hot moulding and heat treatment (post-curing). Dry mixes were prepared using an electric mixer at 3000 rpm for 8 min and preformed under a pressure of 23 MPa. A flash mould method was used for hot moulding. The pressure distributed over a disc pad friction surface during hot moulding is referred to here as the moulding pressure. The five manufacturing parameters (see Table 5.28), i.e. the moulding pressure, the moulding temperature and time, and the heat treatment temperature and time, were varied during manufacturing. As investigated in [171], the results implied that the frictional performance can be strongly affected by the manufacturing conditions. The moulding pressure and temperature can be considered as the dominant manufacturing parameters, according to [171]. That is why the influence of changes in the manufacturing conditions, as shown in Table 5.28, was analysed simultaneously with other influencing factors. The manufacturing parameters used in the production of the disc pad assemblies were different for each of the friction materials A, B, C and D. The moulding pressure used in the manufacture of friction material E was outside the range used for materials A, B, C and D (see Table 5.28 again). The other four manufacturing parameters of friction material E (the moulding temperature and time and the heat treatment temperature and time) used in the test data set were different from those used in the training data set.

An approach to the prediction of the behaviour of friction materials with different formulations and different manufacturing and operating conditions using ANNs was illustrated in Figure 5.64. The main aim was to establish the relationship between the input and output parameter spaces. The total number of input parameters was 26 (18 for the friction material formulation, 5 for the manufacturing parameters and 3 for the testing conditions), and there was one output parameter (the brake factor *C* or the friction coefficient). There is no doubt that the relationship between the input and output spaces is a complex function. Such complexity of the function should be expected in this case, where the predictions are related to the behaviour of friction materials before and after significant thermal load. It is known that when a friction interface experiences severe braking conditions at high temperatures, the friction film includes char and other reaction products such as oxides. That is why the behaviour of a friction material may be seriously changed after the friction interface has experienced a certain thermal history. Extending the capabilities of ANNs to learn about such influences, i.e. to represent differences in the behaviour of a friction material before and after braking at elevated temperatures under the same testing conditions, is extremely important.

Artificial neural network model

Inspired by the biological nervous system, ANN technology has been used to solve a wide variety of complex scientific, engineering and business problems. As has been explained, ANNs are ideally suited to such problems because, like their biological counterparts, they can learn, and therefore can be trained to find solutions, recognize patterns, classify data and forecast future events [172]. According to [173, 174], a neural network is a system composed of many simple processing elements operating in parallel, whose function is determined by the network structure, the connection strengths and the processing performed at the computing elements or nodes. A neural network is a massively parallel distributed processor that has a natural propensity for storing experiential knowledge and making it available for use. It resembles the brain in two important respects: (i) knowledge can be acquired by the network through a learning process, and (ii) interneuron connection strengths, known as synaptic weights, are used to store this knowledge.

The network processes a number of inputs from the outside world to produce an output (i.e. the network's predictions). The neurons are connected by weights and are grouped into layers by their association with the outside world. For example, if a neuron receives data from outside of the network, it is considered to be in the input layer. If a neuron contains the network's predictions, it is in the output layer. The neurons between the input and output layers are in the hidden layer, which serve to (i) add non-linearity to the system and (ii) address interactions between input variables. There are can be many hidden layers, i.e. many levels of non-linearity and many interactions [175]. The number of hidden-layer neurons is one of the most important considerations when solving problems [176].

The implementation of an ANN for prediction of the behaviour of friction materials or, more precisely, prediction of their performance conditioned by the interaction between the friction material and a brake disc, is a complex task. Above all, the architecture of the ANN needs to be properly determined. The architecture of a network is described by how many layers the network has, the number of neurons in each layer, the transfer function of each layer and how the layers are connected to each other. The definition of the architecture of an ANN, according to [177], means determining the numbers of neurons in the input, output and hidden layers, as well as the number of hidden layers. In order to resolve this question, a relation between the numbers of units in the input, hidden and output layers was proposed in [178]. According to [169, 179], the major issue is to determine the numbers of hidden units and weight coefficients. Various methods to set the strengths or weights of the connections between neurons exist. One way is to set the weights explicitly using a priori knowledge. Another way is to train the neural network by feeding it teaching patterns, and the initially set weights are then changed according to a predefined learning rule.

The best architecture to use depends on the type of problem to be represented by the network. A network effects a computation by mapping input values to output values. That is why the ANN technology often requires three sets of data – training data, validation data and test data. The training data set is used to guide the training process, i.e. to update the weights of the neural network during training. The number of training patterns influences the generalization capacity of the network. That number, according to [177, 180], should be several times larger than the capacity of the ANN. The validation data set is used to monitor the quality of the training of the neural-network model, in order to indicate when to terminate the training process. The test data set is used to examine the final quality of the neural model developed.

Backpropagation training, which was used in the study presented here, begins by presenting an input pattern vector to the network, which sweeps forward through the system to generate an output response vector, and the errors at each output are computed. The error signal is the product of the differences between the desired and actual output values. This error signal is then backpropagated to the lower layers in order to find the global minimum of the area of the error surface. It is obvious that neural-network training requires a sufficient number of input and output pairs in order to approximate the relationship between them, but this applies not only to the data which are included in the training set. This process is affected by the representational power of the network and the learning algorithm used [174, 177, 178, 180–183]. The learning ability of neural networks, which extends their prediction power to data outside the training data set, is essential for the successful implementation of ANNs for the prediction of the performance of brake friction materials.

It is clear that sufficient input/target pairs have to be stored in the training data set. The input/output parameters obtained from the formulation, manufacturing and testing of the five types of friction material represented a data set that could be used for training, validation and testing. The four types of materials A, B, C and D were tested to form the training data set. The total number of output results that could be collected from the friction material testing according to the testing methodology adopted (Table 5.26) was 90. This means that 360 output input/output pairs were available for neural-network training. Since the focus was on prediction of the cold performance of brakes, 200 output results were used (50 results per friction material type). Therefore, 150 data points were used for the training of several neural networks with different network topologies and learning algorithms, and 50 data points were used for the validation data set. As mentioned earlier, the testing results for friction material E served as the test data set (50 results).

Thus, the best neural-network architecture and learning algorithm were unknown in advance. A trial and error method had to be employed during the training process to find the network characteristics (the number of hidden layers, the number of processing units (neurons), the values of the connections (weights) between neurons, the learning algorithm etc.) required to match a particular input/output function. Based on MATLAB 6.5 Release 13, the following network architectures were investigated: (i) the one-layered structures 26 $[1]_1$ 1, 26 $[2]_1$ 1, 26 $[3]_1$ 1, 26 $[5]_1$ 1 and 26 $[8]_1$ 1; (ii) the two-layered structures 26 $[1\text{-}1]_2$ 1, 26 $[2\text{-}2]_2$ 1, 26 $[3\text{-}2]_2$ 1, 26 $[5\text{-}2]_2$ 1 and 26 $[8\text{-}2]_2$ 1; and (iii) the three-layered structures 26 $[3\text{-}2\text{-}2]_3$ 1, 26 $[4\text{-}3\text{-}2]_3$ 1, 26 $[4\text{-}2\text{-}2]_3$ 1, 26 $[5\text{-}2\text{-}2]_3$ 1 and 26 $[8\text{-}2\text{-}2]_3$ 1. These network architectures were trained by the Levenberg–Marquardt, Bayesian, resilient backpropagation, scaled conjugate gradient and gradient descent training algorithms. A sigmoid activation function, written as

$$f(x) = \frac{1}{1+e^{-x}}, \quad [5.86]$$

was used between the input and hidden layers.

Pre-processing of the input parameters was done before neural network training. Thus, 18 input parameters related to the formulation of the friction materials were presented to the network as volume percentages, and 5 input manufacturing parameters and 3 input testing parameters were scaled in the range between 0 and 1 according to Eq. 5.87. On the other hand, the output parameter was linearized according to Eq. 5.88. A linear activation function was employed between the hidden and output layers. In order to improve the learning of the ANN and avoid overfitting due to the high-dimensional input space (26 input parameters), the network's outputs were limited to a relatively small range (see Eq. 5.88). Furthermore, the target values were forced into a range around the upper bound of the sigmoid activation function used in the hidden layers.

The following equations were used:

$$I_{\text{scal}} = 1 + \frac{I_{\text{Curr}} - I_{\text{Max}}}{I_{\text{Max}} - I_{\text{Min}}}, \quad [5.87]$$

where I_{Curr} is the current input value, I_{Max} is the maximum input value and I_{Min} is the minimum input value, and

$$O_{\text{Lin}} = 0.75 + 0.001\frac{O_{\text{Curr}} - O_{\text{Min}}}{O_{\text{Max}} - O_{\text{Min}}}, \quad [5.88]$$

where O_{Curr} is the current output value, O_{Max} is the maximum output value and O_{Min} is the minimum output value.

Results and discussion

Different architectures of artificial neural networks and learning algorithms were analysed in order to find the best neural-network architecture for representing the cold performance of brake friction materials under prescribed testing conditions. The networks and learning algorithms analysed, after training and validation, were employed to predict the cold performance of friction material E (the test data set) before and after fading and recovery tests. The quality of the predictions was evaluated by the difference between the predicted and actual values of the brake factor C. Since a disc brake was tested, the brake factor corresponds to twice the value of the friction coefficient in the contact area of the friction pair (i.e. $C=2\mu$). For this purpose, six error intervals were defined (0–5%, 5–10%, 10–15%, 15–20%, 20–25% and 25–30%). Based on the calculated errors between the predicted and the actual values, the numbers of predicted results belonging to each of these error intervals were calculated and expressed as a fraction of the test data set. The numbers of hidden layers and of neurons in the hidden layers in the architectures of the neural networks were varied. The aim of this variation of the architectures adopted was to investigate the influence of the number of hidden layers and of the number of neurons in each of these layers on the prediction capabilities of the network.

The results related to the prediction of the cold performance of a brake using the Bayesian regulation algorithm are shown in Figure 5.65. It can be seen how the different architectures of the neural networks affect the quality of the predictions, using the same input parameters and learning algorithm. The number of neurons in the hidden layer influences the prediction capabilities of the one-layered neural networks in different

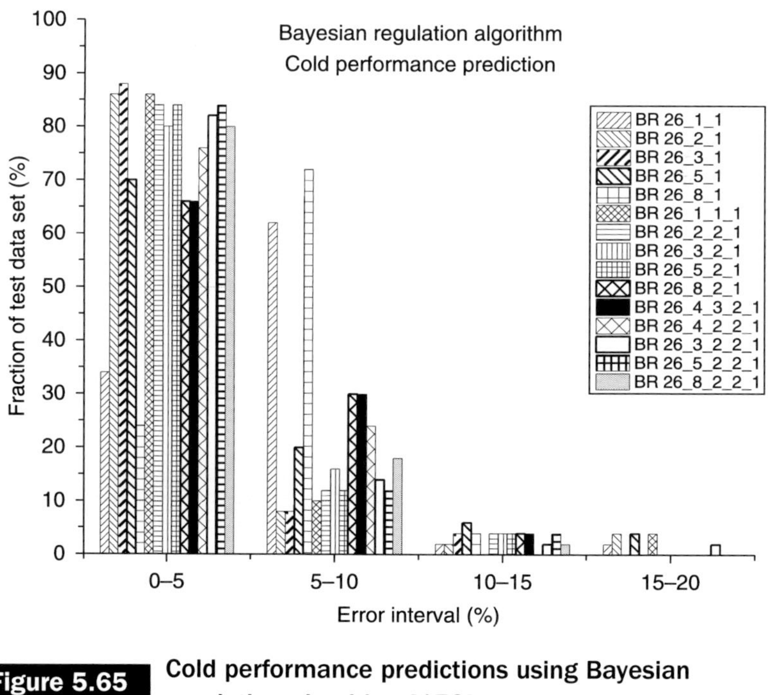

Figure 5.65 **Cold performance predictions using Bayesian regulation algorithm [170]**

ways. Regarding the first error interval (error between 0 and 5%), the best result is shown by the neural network 26 $[3]_1$ 1, where 88% of the predicted results have an error less than or equal to 5%. When the number of neurons in the hidden layers is increased from one to three, the quality of prediction in the first error interval (0–5%) is improved. The quality of prediction deteriorates (only 24% of the results are in this interval for the network 26 $[8]_1$ 1) when the number of neurons is increased to eight (see Figure 5.65). Obviously, the one-layered neural networks trained by the Bayesian algorithm are sensitive to the number of neurons in the hidden layer. In the second error interval (5–10%), the number of predicted results was increased when the number of neurons in the hidden layer was increased (78% for 26 $[8]_1$ 1). On the other hand, the neural network 26 $[3]_1$ 1 had only 8% of its predicted results with an error between 5 and 10%.

In the case of the two-layered neural networks, the number of neurons in the first hidden layer was varied from one to eight while the number of neurons in the second hidden layer was kept constant (two neurons). The two-layered neural-network architectures, under the same initial conditions (inputs and algorithm), showed more stable results over the error intervals. It can be seen from Figure 5.65 that the quality of prediction in the first error interval decreases slowly from 84% (network 26 $[1\text{-}1]_2$ 1) to 66% (network 26 $[8\text{-}2]_2$ 1), except for the network 26 $[5\text{-}2]_2$ 1, as the number of neurons in the first hidden layer is increased. The two-layered neural networks show a relatively low number of predicted results in the second error interval as the number of neurons in the first hidden layer is increased (see Figure 5.65 again). At the same time, the prediction errors in the error interval 10–15% are kept at a constant level as the number of neurons is increased.

The number of neurons in the three-layered neural-network architectures was varied between three and eight in the first hidden layer, while the other two hidden layers were kept at two neurons per layer except in the network 26 $[4\text{-}3\text{-}2]_3$ 1. Regarding the first error interval, these networks show an approximately constant quality of prediction (around 80% except for the network 26 $[4\text{-}3\text{-}2]_3$ 1). Taking into consideration the first and second error intervals, the best prediction is provided by the network 26 $[5\text{-}2\text{-}2]_3$ 1, while the worst result is provided by the network 26 $[4\text{-}3\text{-}2]_3$ 1 (see Figure 5.65). Generally, when the number of neurons in the first hidden layer is increased, the quality of prediction is increased.

The three types of neural networks mentioned (one-, two- and three-layered) were also trained and tested by the resilient backpropagation

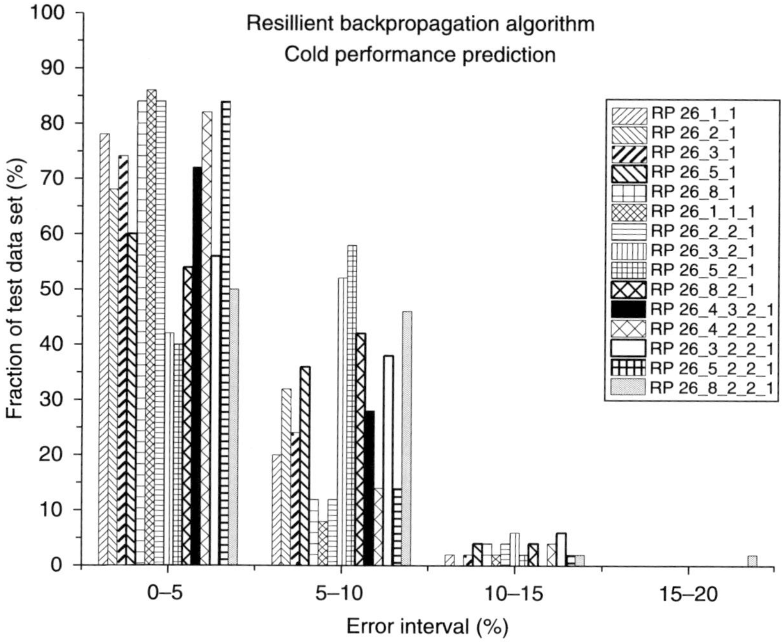

Figure 5.66 **Cold performance predictions using resilient backpropagation algorithm [170]**

algorithm (Figure 5.66), with the same aim. In particular, we investigated how this training algorithm affected the quality of the cold-performance prediction compared with the other training algorithms. For the one-layered neural networks, the quality of prediction in the first error interval (0–5%) is decreased when the number of neurons is increased from one to five, but in the case of the network 26 $[8]_1$ 1 a different trend is found. The number of predicted results with an error between 0 and 5% increases to 84% using the network 26 $[8]_1$ 1. On the other hand, in the second error interval (5–10%), the influence of the number of neurons is inversely related to its influence in the first interval. The two-layered networks 26 $[1\text{-}1]_2$ 1 and 26 $[2\text{-}2]_2$ 1 exhibit high prediction performance for all error intervals. Furthermore, it is evident that, for the two-layered networks, the quality of prediction is decreased when the number of neurons in the first hidden layer is increased further. Therefore, the best results are shown by the three-layered neural network 26 $[5\text{-}2\text{-}2]_3$ 1. In the error interval 0–5%, the quality of prediction is increased when the number of neurons in the first hidden layer is increased from three to five. A further increase in the number of neurons in the first hidden layer causes a significant decrease in the quality of prediction. The neural networks trained by this algorithm are more sensitive to the number of neurons in the hidden layers, particularly in the case of the two- and three-layered neural networks.

The distribution of the predicted results over the six error intervals using the Levenberg–Marquardt algorithm is shown in Figure 5.67. The best result for the one-layered neural networks (84%) is related to the network 26 $[2]_1$ 1. The neural-network architectures 26 $[2\text{-}2]_2$ 1 and 26 $[3\text{-}2]_2$ 1 are the best from the group of two-layered neural networks. Regarding the three-layered networks, the best result in the first error interval is achieved by the network 26 $[4\text{-}2\text{-}2]_3$ 1, with 86% of the predicted results in the first interval. It is evident that the quality of prediction in the first error interval is influenced by increasing the number of neurons in the first hidden layer, in such a way that the number of the predicted results in this interval is decreased. As expected, in the second error interval (5–10%), the number of predicted results is increased when the number of neurons in the first hidden layer is increased. Compared with the other training algorithms analysed so far, it can be seen from the

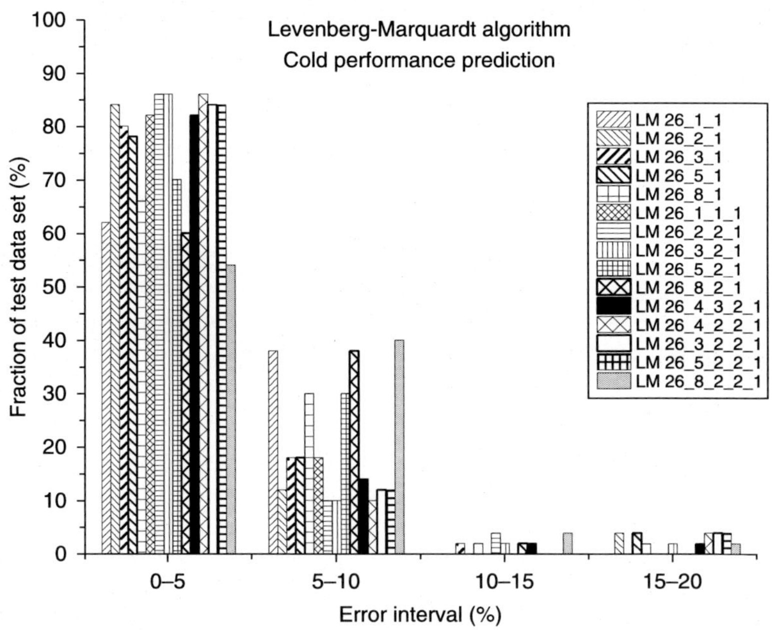

Figure 5.67 **Cold performance predictions using Levenberg–Marquardt algorithm [170]**

results in the last error interval (15–20%) that the three-layered neural networks trained by the Levenberg–Marquardt algorithm show a significant tendency towards overfitting (see Figure 5.67 again).

The one-layered neural networks trained by the scaled conjugate algorithm are less sensitive to an increase in the number of neurons in the hidden layer (according to the results in the first error interval). The best result is achieved by the network 26 $[2]_1$ 1 (Figure 5.68). The quality of the prediction is generally decreased by the introduction of the two-layered neural networks. On the other hand, the best results are provided by the three-layered neural networks (network 26 $[5\text{-}2\text{-}2]_3$ 1). However, the number of predicted results in the first error interval, using the three-layered networks, is drastically decreased when the number of neurons in the hidden layers is increased further. The performance of the neural networks trained by the resilient backpropagation algorithm and the scaled conjugate algorithm is similar from the point of view of the sensitivity of the neural networks to the numbers of hidden layers and neurons.

Finally, the prediction capabilities of the neural networks trained by the gradient descent algorithm were also investigated (Figure 5.69). In

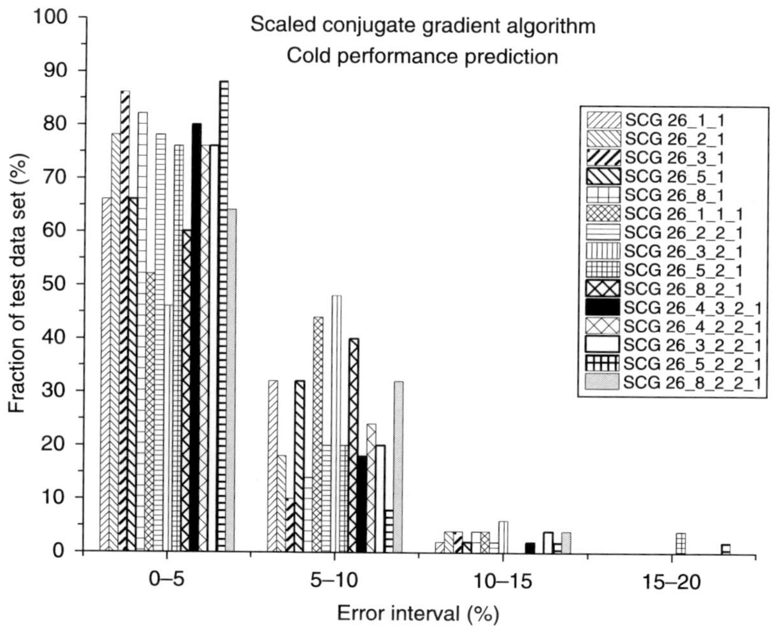

Figure 5.68 Cold performance predictions using scaled conjugate gradient algorithm [170]

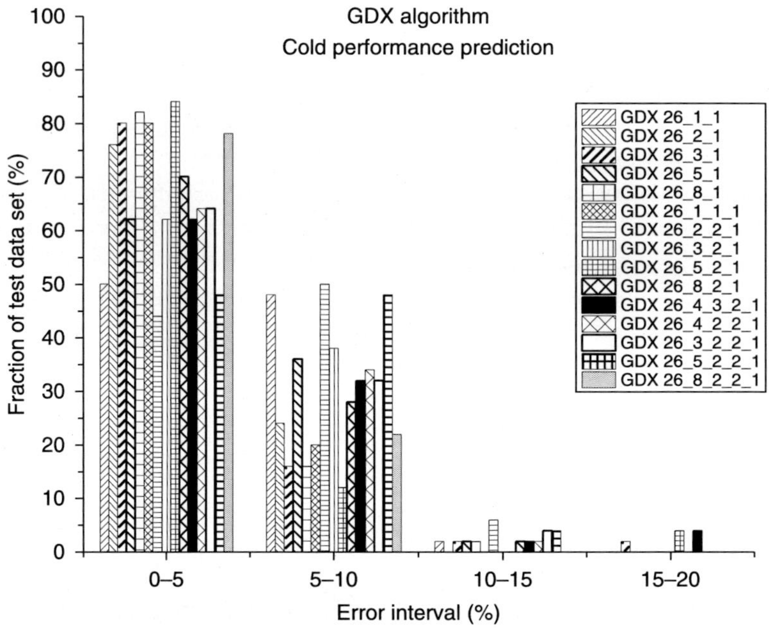

Figure 5.69 **Cold performance predictions by gradient descent algorithm [170]**

the case of the one-layered neural networks, the quality of the prediction is increased by increasing the number of neurons in the hidden layer. The best result is achieved by the network 26 $[8]_1$ 1, where 82% of the predicted results are characterized by an error less than or equal to 5%. The prediction capabilities of the two-layered neural networks are improved by increasing the number of neurons in the first hidden layer. It can be seen in Figure 5.69 that for the neural network 26 $[5\text{-}2]_2$ 1, 84% of the predicted results show an error less than or equal to 5%. A further increase in the number of neurons to eight causes the prediction capability of these networks to deteriorate (70%). In contrast, the best result for the three-layered neural networks is achieved by increasing the number of neurons in the first hidden layer to eight (26 $[8\text{-}2\text{-}2]_3$ 1–78%).

Taking into account the results obtained, based on an analysis of 15 different architectures of ANNs trained by five different learning algorithms, the prediction capabilities of the networks can be summarized. The best five ANNs regarding the prediction results are shown in Table 5.29. It can be seen that different architectures of the neural networks achieve the best results for different training algorithms. It is

Table 5.29 **Fraction of the test data set (%) in the error intervals [170]**

Network	Error interval (%)			
	0–5	5–10	10–15	15–20
BR 26 $[3]_1$ 1	88.0	8.0	4.0	0.0
RP 26 $[5\text{-}2\text{-}2]_3$ 1	84.0	14.0	2.0	0.0
LM 26 $[4\text{-}2\text{-}2]_3$ 1	86.0	10.0	0.0	4.0
SCG 26 $[5\text{-}2\text{-}2]_3$ 1	88.0	8.0	2.0	2.0
GDX 26 $[5\text{-}2]_2$ 1	84.0	12.0	0.0	4.0

evident that the two- and three-layered neural networks with five neurons in the first hidden layer show very good prediction capabilities when trained by different training algorithms. However, the one-layered neural network 26 $[3]_1$ 1 trained by the Bayesian algorithm shows the best results regarding the number of predicted results in the first and second error intervals. An explanation for the results obtained with the Bayesian algorithm may be found in the fact that the result of conventional network training is a single set of weights that can be used to make predictions. In contrast, the result of Bayesian training is a posterior distribution over the network weights. If the inputs of the network are set to new values for a new case, the posterior distribution over the network weights will give rise to a distribution over the output of the network, which is known as the predictive distribution for this new case [184]. The results of prediction by the network 26 $[3]_1$ 1 trained by the Bayesian algorithm will now be analysed further in detail.

As stated earlier (Table 5.26), the 'cold performance 1' test consisted of 25 braking cycles. The brake actuation pressure was between 20 and 100 bar with a uniform distribution (20, 40, 60, 80 and 100 bar). For each pressure value, the initial speed was also uniformly distributed between 20 and 100 km/h. The brake interface temperature was kept at the level of a cold brake ($T < 100\,°C$). These initial conditions were chosen in order to investigate the sensitivity of the friction materials to pressure and speed changes. Accordingly, it was interesting to develop a neural model able to learn and generalize these influences. Therefore, the network 26 $[3]_1$ 1 trained by the Bayesian algorithm was employed to predict the influence of the pressure and initial speed for known input parameters. The input parameters were defined by the formulation and the manufacturing conditions of the friction material E stored in the test data set (see Tables 5.27 and 5.28). In order to illustrate the ability of the neural network 26 $[3]_1$ 1 to predict the characteristics of the brake friction material in the 'cold performance 1' test, the real and predicted minimum and maximum values of the brake factor *C* are shown in Table 5.30. The differences between the real and predicted results for the brake factor (minimum and maximum values) were calculated; see Tables 5.30 and 5.31 (where $\Delta C = C_{Max} - C_{Min}$). These values can be also seen in Figures 5.70–5.75.

Table 5.30 Speed sensitivity – comparison between maximum and minimum values (real and predicted) (cold performance 1, $v_{Initial} = 20–100$ km/h) [170]

p (bar)	Minimum brake factor *C*		Maximum brake factor *C*		$\Delta C = C_{Max} - C_{Min}$	
	Real	Predicted	Real	Predicted	Real	Predicted
20	0.883	0.914	0.933	0.94	0.05	0.026
40	0.924	0.914	0.986	0.935	0.062	0.021
60	0.934	0.9	0.973	0.94	0.039	0.04
80	0.89	0.872	0.931	0.895	0.041	0.023
100	0.842	0.855	0.906	0.875	0.06	0.02

Table 5.31 Pressure sensitivity – comparison between maximum and minimum values (real and predicted) (cold performance 1, $p = 20–100$ bar) [170]

v (km/h)	Minimum brake factor *C*		Maximum brake factor *C*		$\Delta C = C_{Max} - C_{Min}$	
	Real	Predicted	Real	Predicted	Real	Predicted
20	0.906	0.875	0.986	0.94	0.08	0.065
40	0.89	0.872	0.961	0.931	0.071	0.059
60	0.88	0.864	0.94	0.92	0.06	0.056
80	0.873	0.862	0.937	0.914	0.064	0.052
100	0.849	0.855	0.934	0.92	0.085	0.065

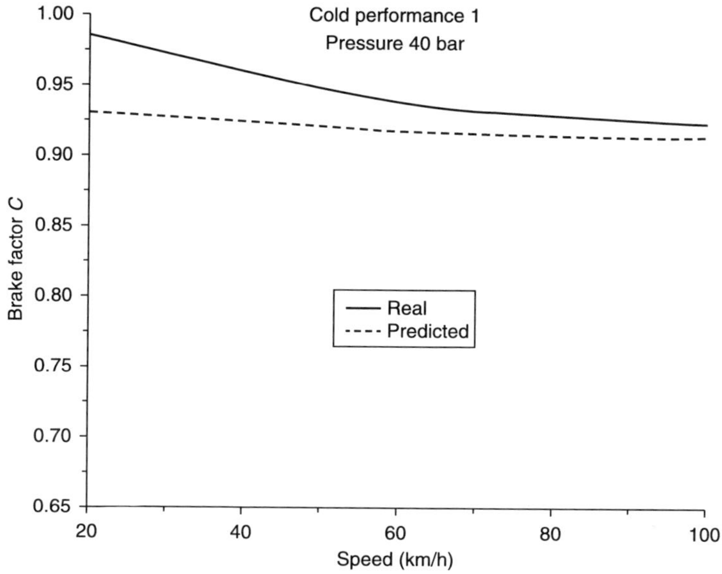

Figure 5.70 **Cold performance 1 – comparison between real and predicted results versus initial speed for $p=40$ bar [170]**

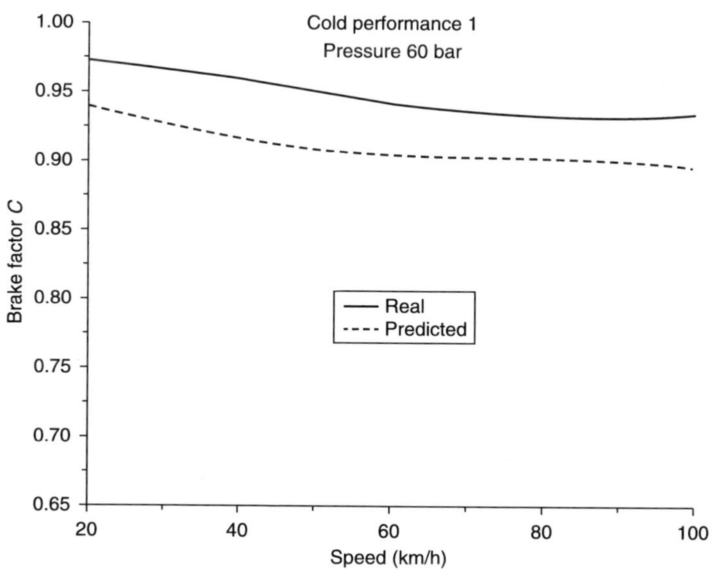

Figure 5.71 **Cold performance 1 – comparison between real and predicted results versus initial speed for $p=60$ bar [170]**

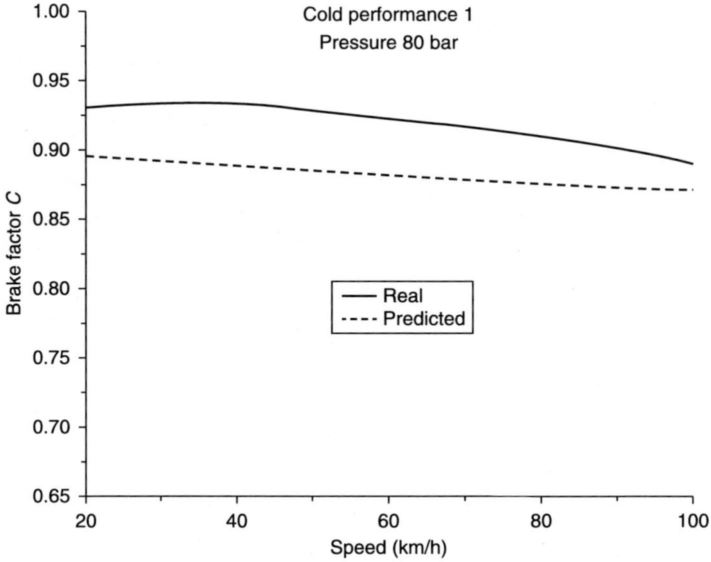

Figure 5.72 **Cold performance 1 – comparison between real and predicted results versus initial speed for *p*=80 bar [170]**

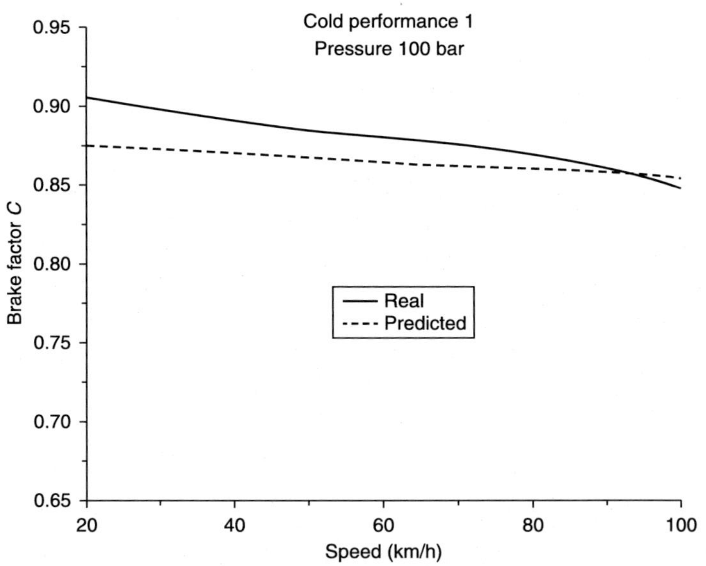

Figure 5.73 **Cold performance 1 – comparison between real and predicted results versus initial speed for *p*=100 bar [170]**

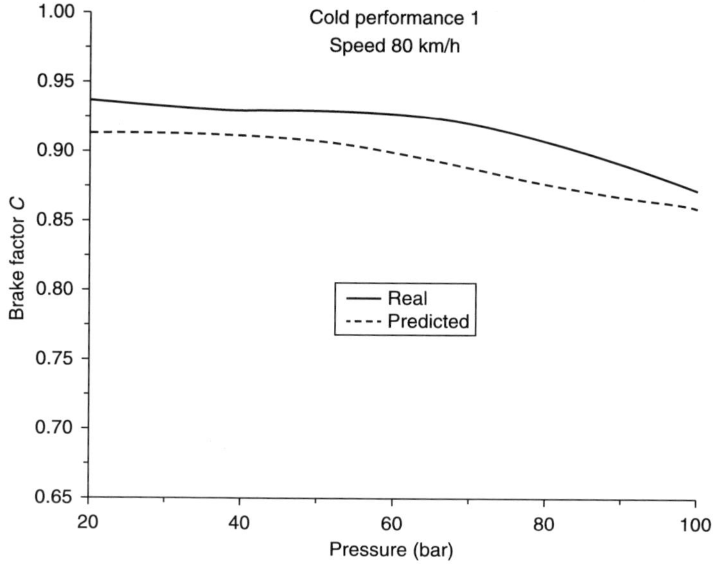

Figure 5.74 Cold performance 1 – comparison between real and predicted results versus brake actuation pressure for $v = 80$ km/h [170]

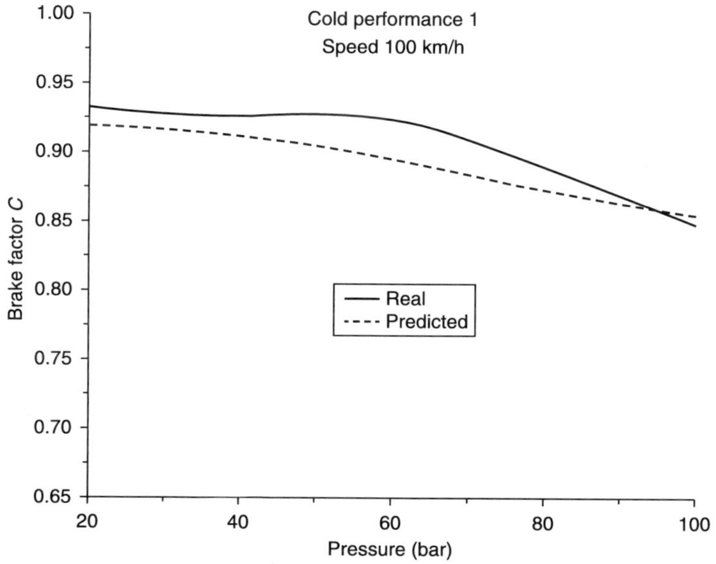

Figure 5.75 Cold performance 1 – comparison between real and predicted results versus brake actuation pressure for $v = 100$ km/h [170]

The 'cold performance 1' test offered interesting possibilities to investigate the influence of the brake actuation pressure on the brake performance for different initial speeds and vice versa. An estimation of the ability of the neural network to learn and generalize this influence, in the case of friction material E, is presented in Figures 5.71–5.76 and Tables 5.30 and 5.31. From these figures, it can be seen that the real brake performance is decreased when the brake actuation pressure and initial speed are increased. The neural model recognizes this behaviour of the friction material. Furthermore, it is evident an increase in the brake actuation pressure up to 60 bar is accompanied by relatively stable brake performance. An increase in the brake actuation pressure over 60 bar causes a decrease in the brake performance. From Figures 5.74 and 5.75, it can also be seen that there is a non-linearity in the real variation of the brake factor *C* due to the sensitivity of the friction material E to pressure changes. However, it can be concluded that the predictions of the one-layered neural network 26 $[3]_1$ 1 are more linear owing to the one-layered network architecture and the fact that it has only three neurons in the hidden layer. This type of neural network is able to approximate non-

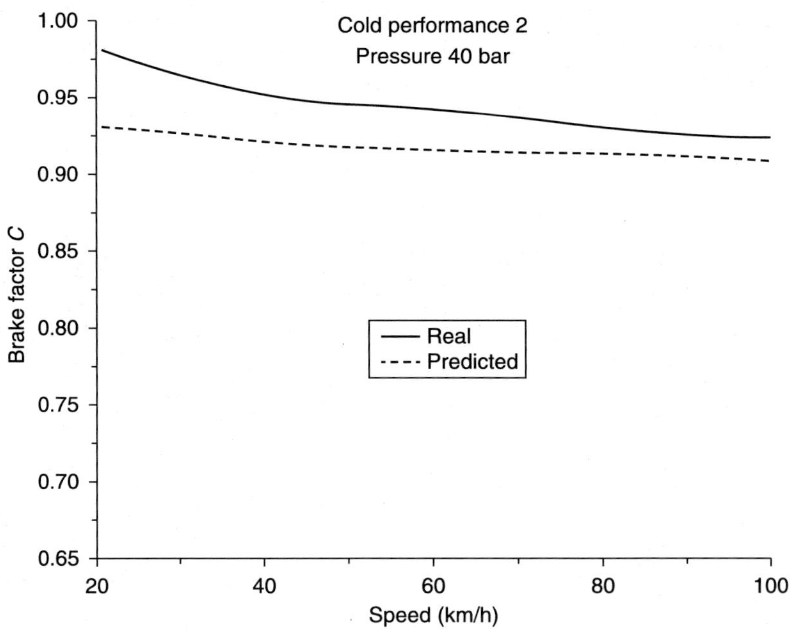

Figure 5.76 **Cold performance 2 – comparison between real and predicted results versus initial speed for $p = 40$ bar [170]**

linear functions of the brake factor C. It should be taken into consideration that a very high number of input parameters was processed in order to approximate the complex functional relationship between the inputs and the output. That is why a neural-network model with a limited number of degrees of freedom was chosen.

It has previously been mentioned that the cold performance of the brake friction material could change dramatically before and after the fading and recovery tests. In order to test the capability of the ANNs to learn these influences, the training of the ANNs was performed in such a way that data related to the 'cold performance 2' test was fed to the neural network after data related to the fading and recovery tests. In this case, the fading test was performed in accordance with the testing methodology presented in Table 5.26. The maximum temperature reached after 15 braking applications was 310 °C, while the temperature reached after 3 min cooling at 300 °/min was 240 °C. At this temperature, the recovery test was performed according to the conditions shown in Table 5.26. At the end of the recovery test, the brake was cooled until its temperature, measured by a thermocouple, dropped below 100 °C. After that, the cold brake was tested in the same way as in the 'cold performance 1' test. Differently from the 'cold performance 1' test, this test was therefore done on a brake that had experienced a certain thermal history. The results related to prediction of the results of the 'cold performance 2' test for the friction material E are shown in Figures 5.75 and 5.76, and in Tables 5.32 and 5.33.

Table 5.32 Speed sensitivity 2 – comparison between maximum and minimum values (real and predicted) (cold performance 2, $v_{Initial}$ = 20–100 km/h) [170]

p (bar)	Minimum brake factor *C*		Maximum brake factor *C*		$\Delta C = C_{Max} - C_{Min}$	
	Real	Predicted	Real	Predicted	Real	Predicted
20	0.873	0.914	0.98	0.94	0.107	0.026
40	0.924	0.909	0.98	0.931	0.056	0.022
60	0.908	0.892	0.961	0.917	0.053	0.025
80	0.856	0.867	0.935	0.897	0.079	0.03
100	0.863	0.855	0.903	0.878	0.04	0.023

Table 5.33 Pressure sensitivity 2 – comparison between maximum and minimum values (real and predicted) (cold performance 2, $p = 20–100$ bar) [170]

v (km/h)	Minimum brake factor C		Maximum brake factor C		$\Delta C = C_{Max} - C_{Min}$	
	Real	Predicted	Real	Predicted	Real	Predicted
20	0.873	0.878	0.98	0.942	0.107	0.064
40	0.828	0.869	0.947	0.928	0.119	0.059
60	0.896	0.864	0.944	0.917	0.048	0.053
80	0.893	0.861	0.929	0.917	0.036	0.056
100	0.856	0.855	0.924	0.914	0.068	0.059

Regarding the real and predicted results of the 'cold performance 2' test, an evident influence of the thermal history can be seen (Figures 5.76–5.81). Taking into consideration the differences ΔC between the maximum and minimum values of the brake factor C as functions of the initial speed and the pressure (see Tables 5.32 and 5.33, respectively), the

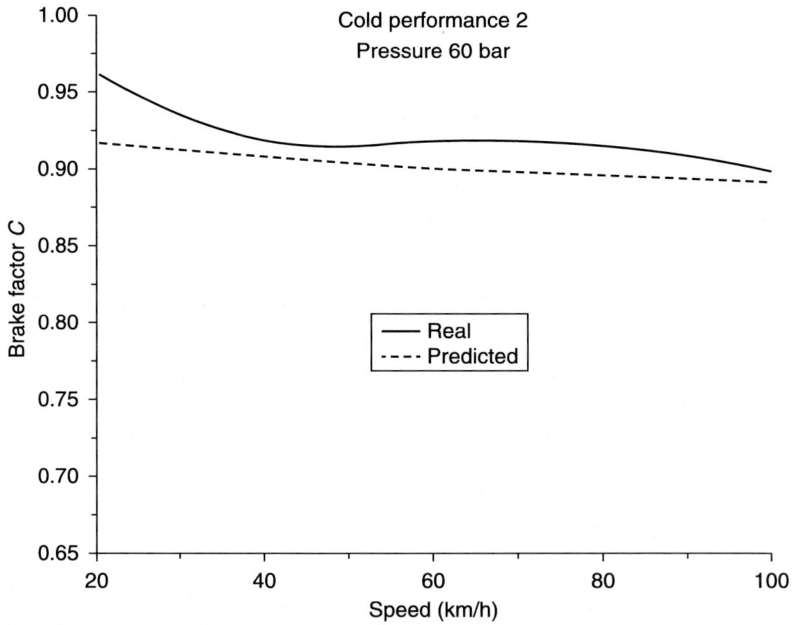

Figure 5.77 Cold performance 2 – comparison between real and predicted results versus initial speed for $p = 60$ bar [170]

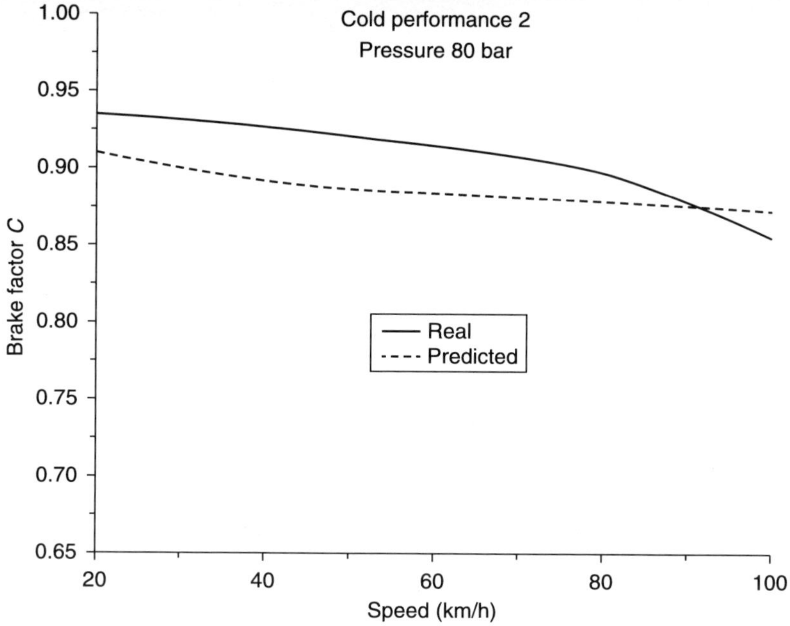

Figure 5.78 Cold performance 2 – comparison between real and predicted results versus initial speed for $p=80$ bar [170]

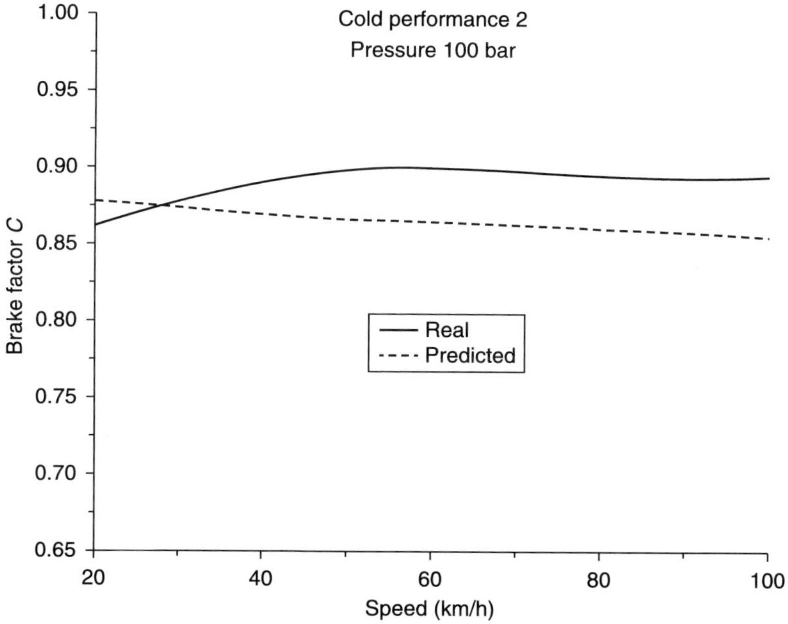

Figure 5.79 Cold performance 2 – comparison between real and predicted results versus initial speed for $p=100$ bar [170]

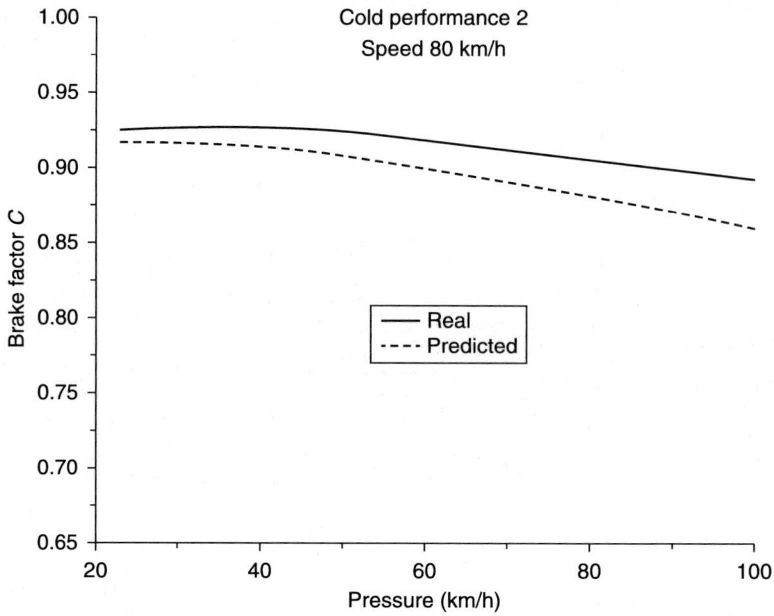

Figure 5.80 **Cold performance 2 – comparison between real and predicted results versus brake actuation pressure for $v = 80$ km/h [170]**

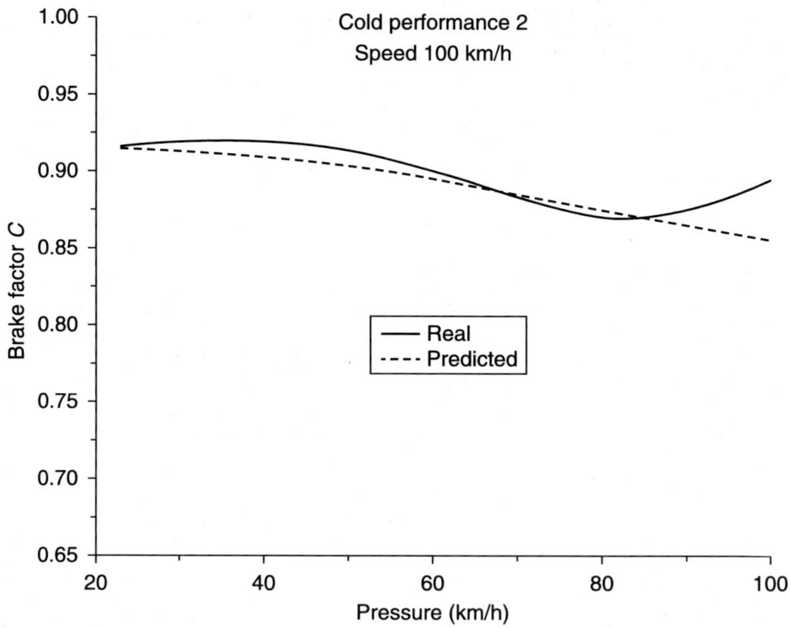

Figure 5.81 **Cold performance 2 – comparison between real and predicted results versus brake actuation pressure for $v = 100$ km/h [170]**

variations of the brake factor in the 'cold performance 2' test with changes in both the initial speed and, particularly, the brake actuation pressure are higher than in the 'cold performance 1' test. However, the capabilities of the neural network to generalize the influence of the thermal history are at an acceptable level.

5.3.2 Modelling and prediction of the fade performance of brake friction materials

Introduction

Competitive advantages can be achieved in the friction materials industry by appropriate management of the formulation and manufacturing conditions of friction materials, so that changes can be made and implemented skilfully in a cost-effective manner. The management of such changes has a decisive influence on whether a new friction material can be launched onto the market. The basic requirements imposed on friction materials are related to the level of the friction coefficient and its stability in different operating regimes. Thus, brake performance is determined mostly by the behaviour of the friction material. The performance of friction materials should be characterized by a relatively high, stable friction force, reliable strength and good wear resistance, irrespective of temperature, humidity, age, the degree of wear and corrosion, the presence of dirt, water sprayed from the road, etc. Accordingly, the development of friction materials and the tailoring of their performance is becoming a demanding task. Many formulations of brake friction materials have to be evaluated. It is well known that the sensitivity of the friction coefficient to the operating conditions, i.e. the sliding speed, the brake actuation pressure and the brake interface temperature, is substantially influenced by the situation in the contact area of the brake friction pair. The contact situation between a cast iron brake disc and an organic brake pad is complicated and can vary dramatically with different material combinations. This is a result of the wide diversity in properties (mechanical and chemical) of the constituents of brake pads [185, 186].

To understand the development of friction materials, it is necessary to understand the interfacial relations between the rubbing surfaces. There are many asperities between the rubbing surfaces, leading to the formation of many micro-contact points. These points change dynamically from place to place in fractions of a second during rubbing [187]. The total contact area is unknown and depends on the interactions in the friction pair. Furthermore, it is known that when two bodies slide against each other with a relative speed and a contact pressure, frictional heat is generated at the sliding interface. The subsequent thermomechanical deformation of the bodies modifies the contact profile and the pressure distribution, altering the amount of frictional heat generated. The frictional heat generated during brake application can easily raise the temperature at the friction interface above the glass transition temperature of the binder resin, and often above the decomposition temperature. Evolution of gas at the braking interface because of pyrolysis and thermal degradation of the material results in a decrease in the friction force at elevated temperatures. The thermal load in the contact region of a brake pad does not build up homogeneously, and hot rings can be often seen on a brake disc, which change their radius. Instability of the friction coefficient after a certain number of brake applications is common and depends on the contact interface situation. In general, the variation of the friction coefficient during sliding depends on changes in the real area of contact at the friction interface, in the strength of the binder resin and in the frictional properties of the ingredients at elevated temperatures, and on the formation of a friction film [188].

The friction film, or third-body layer, is produced incessantly, maintaining a certain thickness, and is composed of carbonaceous reaction products, unreacted constituents, oxides from metallic ingredients and other substances [163, 189]. The frictional characteristics of the transfer layer between the brake disc and the pads developed during braking influence the effectiveness of braking at elevated temperatures. This important feature of friction materials is referred to as 'fade' and is of decisive importance in the performance of brake friction materials. The performance of brake friction materials is generally classified by considering five different responses: (i) cold performance 1 (before braking at elevated temperatures), (ii) fading performance (the behaviour of the friction material during subsequent braking, i.e. at elevated temperatures), (iii) recovery performance (braking at elevated temperatures after a period of cooling), (iv) cold performance 2 (after braking at elevated temperatures) and (v) wear behaviour in different temperature regimes. It is obvious that friction materials are expected to satisfy some very complex and mutually opposing requirements. That is why the prediction of the performance of friction materials in relation to effects related to their formulation, manufacturing and operating conditions can help in the development of friction materials.

There is no doubt that changes in engineering are a fact [190]. To support future decision-making related to the influencing factors mentioned above, a new development policy has to be introduced in which the moment of decision-making about the performance of friction materials is changed. This chapter investigates how artificial intelligence techniques can assist in the development of innovative friction materials, using, in particular artificial neural networks. Accordingly, the possibilities for predicting the fading performance of friction materials, regarding the influence of their formulation and manufacturing conditions, were investigated by means of artificial neural networks. Neural modelling of the performance of brake friction materials at elevated temperatures was investigated using two different training data sets for training the neural networks. The difference between these training data sets was related to the size of the sets, the data distribution and the type of data stored. The first training data set consisted of 360 randomly distributed data points related to cold, fading and recovery performance. The main goal of this training data set was to investigate the possibilities for developing a neural model of the behaviour of friction materials capable of predicting the fading performance but at the same time able to predict the behaviour under testing conditions related to cold and recovery performance. The second training data set was obtained by formulating, manufacturing and testing eight different brake friction materials (120 data points). This training data set was used to develop a neural model that would be used only for predicting the fading performance of friction materials.

Modelling of fading performance

The temperature sensitivity of brake friction materials has always been a critical aspect of ensuring their smooth and reliable functioning. This is particularly related to the front brakes, which absorb a major amount (up to 80%) of the total kinetic energy of the vehicle. This kinetic energy is converted into heat, causing the disc to reach high temperatures of up to 400 °C. The severity of such temperature rises is manifested further in the form of very high flash temperatures (600–800 °C) at the contacting asperities. At such high temperatures, the friction force suffers from a loss of effectiveness called μ-fade [188, 191]. This loss of effectiveness cannot be easily predicted, owing to the use of different ingredients in friction materials, different formulations and different choices of manufacturing parameters. It is known that organic constituents, fibrous materials and solid lubricants play important roles in establishing the transfer layer at the friction interface [155, 191]. However, a clear connection between the mixture of components, the manufacturing conditions used and the friction layer is not yet well established, nor has a connection been established between the friction layer and the friction behaviour of the system [163]. There is no doubt that measures have to be taken to optimize the fade performance of friction materials, but trial and error methods need to be eliminated because they are costly and time-consuming. On the other hand, brake pads usually contain more than 20 different ingredients, and because of the existence of a highly complex tribological surface layer, the frictional behaviour of the system can be modulated in different ways.

It has been mentioned that the sensitivity of the fade performance of friction materials to different operating conditions due to the instantaneously established contact situation depends on the types of ingredients selected, the formulation of the friction material and the manufacturing conditions. It should be obvious that the achievement of a stable friction coefficient under severe thermal load, for braking systems with different characteristics, requires engineering changes to be implemented regarding the selection of ingredients, the formulation and the choice of manufacturing conditions. Prediction of the sensitivity of friction materials to temperature changes is critically important because the main constituent, for example phenolic resin, changes its properties above the glass transition temperature. Thermal decomposition of the binder resin provokes changes in the reinforcing capabilities of the fibres used, and that is why a reduction in the friction coefficient (fading) happens. These changes have to be the subject of a systems engineering investigation, and that is why an artificial neural network technology was employed for predicting the behaviour of friction materials under high-temperature load.

Owing to the complex formulation of brake friction materials and the stochastic nature of braking phenomena, artificial intelligence abilities need to be embedded into the process of friction materials development. Generally, intelligence can be defined as the ability to learn and understand, to solve problems, and to make decisions [180]. That is why brain-like information processing is needed. An artificial intelligence technology could introduce a new approach to the development of brake friction materials. When the rules are not known in a situation with fuzzy or incomplete information, a neural-network technology can be applied. The abilities of neural networks have been applied to search numerical data for fuzzy rules, i.e. to 'transform' known data and experience into rules. This is possible because neural networks are capable of handling a large set of complicated information, and they have a reputation for being able to deal with noisy data. That is why artificial neural networks were good candidates for the application considered here, owing to their capabilities for non-linear behaviour, learning from experimental data and generalization. In order to learn a complex input/output functional approximation, artificial neural networks need to be trained so that a particular input leads to a specific target output [169, 192]. It is clear that some experimental data are needed in order that the input/output relationship may be generalized. Therefore, a neural model that is going to be used for predicting the performance of friction materials has to be taught about the specific influences of the inputs on the outputs. The preliminary step in the development of a neural model is the identification of the model's inputs and outputs. Input/output identification depends on the objectives of the model and the choice of the data generator. According to the present objective of modelling and predicting the fading performance of brake friction materials, the input parameters were determined by the formulation and the manufacturing and testing parameters versus changes in the brake factor C (a one-dimensional output parameter). A single-ended full-scale inertia dynamometer was used as a data generator.

The performance of a brake results from complex, interrelated tribo-processes occurring during braking. These tribo-processes, under specified testing conditions, are mostly affected by the physicochemical properties of the ingredients of the friction material. In order to establish relationships between the input and output parameter spaces, the role of the data generator is important from the point of view of repeatability of the testing conditions. That is why it was decided to perform testing of friction materials under strictly controlled conditions related to changes in the brake actuation pressure, the initial speed, the brake interface temperature and the inertia of the revolving masses. These testing conditions were chosen in order to allow us to simulate real operating regimes with a full-scale inertia dynamometer. The dynamometer was equipped with a PC-based automatic control and data acquisition system that measured the pressure, speed, temperature and braking torque at a sampling rate of 50 Hz. Thus the braking torque and pressure could be measured during a braking cycle, and an average value of the brake factor C was calculated, taking account of the known values of the piston diameter and the effective brake disc radius. This calculation was done for speeds during a braking cycle in the range between $0.8v$ and $0.1v$.

Once the range of input parameters had been decided, the next step was to choose a sampling strategy. The input parameters were defined by three different groups of data (the formulation of the friction material, the manufacturing conditions and the testing conditions). The data stored for each of these groups needed to be ranged, and a sampling strategy had to be determined. The most complicated problem was related to the selection of the range of the formulation data. The problem related to the range and distribution of the formulation parameters was complicated by the fact that 150 different raw materials are used in the friction material industry today. Furthermore, the selection of ingredients and the ranges of their proportions depended on the type of friction material (metallic, semi-metallic or non-asbestos organic). Similarly, the selection of the input parameters and ranges for the manufacturing conditions depended also on the type of friction material whose performance was going to be modelled. Regarding the group of testing conditions, that task was much easier: a friction material mounted on a specified brake can be tested according to a chosen testing methodology. The testing methodology needed to be chosen according to the previously selected ranges and distribution of the input/output data. The friction materials were tested according to the methodology shown in Table 5.34.

In order to create the training, validation and test data sets, the ingredients in the friction material formulation were selected and the ranges of the proportions of each ingredient were chosen, as listed in Table 5.35; 18 different ingredients were used. The friction materials used to generate the input/output data were divided into three groups. The first group of friction materials (F1–F4) was used to create the first training set for developing the neural model. The second group (F5–F8)

Table 5.34 **Fading performance – testing methodology [191]**

Test No.	Test	Pressure (bar)	Initial speed (km/h)	Temperature (°C)	Number of braking cycles
1	Burnishing	40	90	<100	100
2	Cold performance 1	20, 40, 60, 80, 100	20, 40, 60, 80, 100	<100	25
3	Fading	Corresponding to 3 m/s^2 deceleration in first braking cycle	90	Open	15 (with 45 s pause between each brake application)

Table 5.35 Friction materials F1–F8; F_{T1} and F_{T2} – ranges of proportions of ingredients used (vol%) [191]

Ingredient	Friction materials F1–F4 (training data set)	Friction materials F5–F8 (training data set), F9 (validation data set)	Friction material F_{T1} test data set	Friction material F_{T2} test data set
Phenolic resin	17–20	17–22		
Iron oxide	3–5	3–7	2	3
Baryte	26–15	26–12	12	18
Calcium carbonate	1–3	1–3		
Brass chips	1–2	1–3	3	3
Aramid	4–6	3–6	7	4
Mineral fibre	12–16	12–14	10	12
Vermiculite	4–7	4–8		
Steel fibre	4–2	4–1	1.5	2
Glass fibre	2–3	2–4	4	3
Brass powder	1–2	1–2	0.8	1
Copper powder	1–2	1–3		
Graphite	7–4	8–3		
Friction dust	5–3	5–2	6	5
Molybdenum disulphide	1–3	1–3		
Aluminium oxide	2–3	2–3	3.5	3
Silica	1–2	1–2		
Magnesium oxide	8–2	8–2		

was used together with the first group (F1–F4) to create the second training set, for a neural model specifically for prediction of fading performance. Besides the friction materials F1–F8, Table 5.35 lists the compositions of two more friction materials, F_{T1} and F_{T2}. These two types of friction material were used for testing the neural model. This means that the neural models were tested using the same test data set but the ratios between the numbers of data points used for training and testing were different. Input/output data related to the friction material denoted by F9 were used for creation of the validation data set.

Therefore, in order to evaluate the generalization capabilities of the trained artificial neural networks, the test data sets were sampled by testing new types of friction material with a formulation different from those used for the training data sets, as shown in Table 5.35. A similar strategy was used for selecting the ranges and distribution of the manufacturing parameters, as shown in Table 5.36. It is evident from Table 5.36 that five manufacturing parameters were considered as input parameters. Completely different manufacturing conditions were employed to realize the friction materials used for testing the neural networks compared with those used for the friction materials F1–F9. Moreover, the manufacturing parameters for the two groups of materials (F1–F4 and F5–F9) were not the same. The ranges of moulding pressure, moulding temperature and moulding time used for manufacturing the materials F5–F9 differed from those used for manufacturing the first group (F1–F4).

The friction materials, which were used as a part of a disc pad assembly, were manufactured by a conventional procedure, consisting of dry mixing, preforming, hot moulding and heat treatment (post-curing). Dry mixes were prepared using an electric mixer at 3000 rpm for 8 min and preformed under a moulding pressure of 23 MPa. The disc pads were designed to be mounted on the front axle of a passenger car, as described in Section 5.3.1.

Table 5.36 Ranges of manufacturing parameters [191]

Manufacturing parameter	Friction materials F1–F4 (training data set)	Friction materials F5–F8 (training data set), F9 (validation data set)	Friction material F_{T1} (test data set)	Friction material F_{T2} (test data set)
Moulding pressure (MPa)	45–65	51–70	40	45
Moulding temperature (°C)	155–170	150–175	170	165
Moulding time (min)	6–11	5–10	10	6
Heat treatment temperature (°C)	200–250	200–250	200	200
Heat treatment time (h)	12–5	12–5	8	5

In the study presented in this section, two sets of training data were created. The results obtained during testing of the first group of friction materials (F1–F4) were used for training the neural networks. In the testing methodology (see Tables 5.26 and 5.34) adopted for the fading test, 15 results were obtained per friction material tested. Accordingly, 60 results for the fading performance could be collected for the friction materials F1–F4. These 60 results related to the fading performance of the materials F1–F4, together with 100 results from the 'cold performance 1' test, 100 results from the recovery performance test and 100 results from the 'cold performance 2' test mentioned earlier, were used for training the artificial neural networks.

In order to investigate the influence of the number of points in the training data, their distribution and the type of data stored in the training data set, the second training data set was created. This set of training data was composed of data on the formulation, manufacturing and testing (of only the fade performance) of the eight types of friction materials F1–F8. Accordingly, 120 results related to the fading performance of the friction materials were collected. These 120 data points were not randomly distributed over the training data set. The training data set was created by arranging groups of input/output data for each friction material in sequence according to the number of brake applications (from the first to the 15th). Consequently, the training data set was defined first by 15 input/output data points related to the friction material F1, followed by 15 input/output data points related to the friction material F2, and finally by 15 input/output data points related to the friction material F8 (i.e. eight groups, F1–F8). This training data set, composed in this way, was expected to cause the minimum disturbance to the training of the neural network [193]. Artificial neural networks trained using this training data set were used to develop a neural model able to predict only the fading performance of friction materials. The training process based on both of the training data sets was validated by a validation data set defined by the input/output parameters obtained during testing of the friction material F9.

The prediction capabilities of the neural networks were tested using test data sets obtained by formulating, manufacturing and testing the friction materials F_{T1} and F_{T2} (30 data points). The materials F_{T1} and F_{T2} for the test data sets were formulated and manufactured differently. The test data set for F_{T1} was created from data obtained after testing of a new type of friction material: 10 of the ingredients used in the formulation of this friction material were outside the ranges used in the training data set obtained from F1–F4.

After generation of the training and test data sets, the training process of the selected artificial neural networks was performed with several different training algorithms. Finding the characteristics of the artificial neural network that provides the best matching between the input and output data stored in the training data set is an important step. Training of neural networks is used to determine the following characteristics of the network: (i) the number of hidden layers, (ii) the number of processing units (neurons) in each hidden layer and (iii) the values of the connections (weights) between the neurons. The best architecture and learning algorithm are unknown in advance. Based on MATLAB 6.5 Release 13, one-layered, two-layered and three-layered network architectures were investigated, as in Section 5.3.1. The architectures of the one-layered networks consisted of only one hidden layer, and the number of neurons in the hidden layer was varied from 1 to 8, giving architectures denoted by 26 $[1]_1$ 1, 26 $[2]_1$ 1, 26 $[3]_1$ 1, 26 $[5]_1$ 1 and 26 $[8]_1$ 1. The structures of the two-layered networks were defined by two hidden layers, and the numbers of neurons in the hidden layers were varied as follows: 26 $[1\text{-}1]_2$ 1, 26 $[2\text{-}2]_2$ 1, 26 $[3\text{-}2]_2$ 1, 26 $[5\text{-}2]_2$ 1, 26 $[8\text{-}2]_2$ 1, 26 $[8\text{-}4]_2$ 1 and 26 $[10\text{-}5]_2$ 1. The architectures of the three-layered networks were defined by three hidden layers, and the numbers of neurons in the hidden layers were varied, except that in the third layer, which was kept at two neurons: 26 $[3\text{-}2\text{-}2]_3$ 1, 26 $[4\text{-}3\text{-}2]_3$ 1, 26 $[4\text{-}2\text{-}2]_3$ 1, 26 $[5\text{-}2\text{-}2]_3$ 1, 26 $[8\text{-}2\text{-}2]_3$ 1 and 26 $[8\text{-}4\text{-}2]_3$ 1. These network architectures were trained using the Levenberg–Marquardt, Bayesian regulation, resilient backpropagation, scaled conjugate gradient and gradient descent training algorithms. In addition, the activation or transfer functions between the input and hidden layers, between the hidden layers, and between the hidden and output layers needed to be chosen. A sigmoid activation function was used between the input and hidden layers and between the hidden layers (see Eq. 5.89 below). This activation function was chosen because it combines nearly linear behaviour, curvilinear behaviour and nearly constant behaviour depending on the value of the input. The sigmoid function, sometimes called the squashing function, takes any real-valued input and returns an output bounded between 0 and 1. In the input range between –1 and 1, the behaviour of the output is nearly linear. As the input is increased, the output of the sigmoid activation function becomes curvilinear. A linear activation function was employed between the hidden and output layers. Pre-processing of the input data was done as described in Section 5.3.1 (Eqs 5.87 and 5.88). The output parameters were linearized. The linearization of the output parameters influences the ability of a neural network to match the input/output relationship. For this reason, the outputs were linearized using Eqs 5.89 and 5.90. Equation 5.89 was used to linearize the outputs for the first group of friction materials F1–F4. In order to avoid overfitting due to the high-dimensional input space (26 input parameters) and the relatively small number of points in the training data, as well as the random data distribution used, the output of the network was limited to a relatively small range. On the other hand, for the second training data set F1–F8 (for fading performance only), the output linearization was changed in order to improve the sensitivity of the network to changes in the inputs. For this reason, the outputs were linearized according to Eq. 5.90. The outputs were linearized in such a way that they were forced into the non-linear (curved) range of the sigmoid activation function. Therefore, owing to the very complex changes in the brake factor *C* during the fading test, the ability of the neural model to predict the fading performance would be significantly improved by introducing a more flexible way to linearize the network's outputs. The equations used were

$$O_{\text{Lin}} = 0.75 + 0.001 \times \frac{(O_{\text{Curr}} - O_{\text{Min}})}{(O_{\text{Max}} - O_{\text{Min}})}, \qquad [5.89]$$

$$O_{\text{Lin}} = 0.75 + 0.2 \times \frac{(O_{\text{Curr}} - O_{\text{Min}})}{(O_{\text{Max}} - O_{\text{Min}})}, \qquad [5.90]$$

where O_{Curr} is the current output value, O_{Max} is the maximum output value and O_{Min} is the minimum output value.

Prediction of fading performance

Based on the numbers of artificial neural networks and training algorithms used, 90 different neural models were developed in order to be tested regarding their abilities for prediction of fading performance. The neural models were tested with new data sets (test data sets) created by formulation, manufacturing and testing of the friction materials F_{T1} and F_{T2}. The input/output parameters of these friction materials were completely different from those used for training and validation. The real fading performance of the friction material F_{T1} used for testing the neural models is shown in Figure 5.82. From Figure 5.82, it can be seen that the changes in the brake factor *C* versus temperature during 15 braking cycles are very complex. It is evident that the material F_{T1} shows substantial fading, which is primarily due to thermal decomposition of the binder resin and other organic constituents, as well as gas formation. Regarding the changes in the brake factor versus temperature, the fading

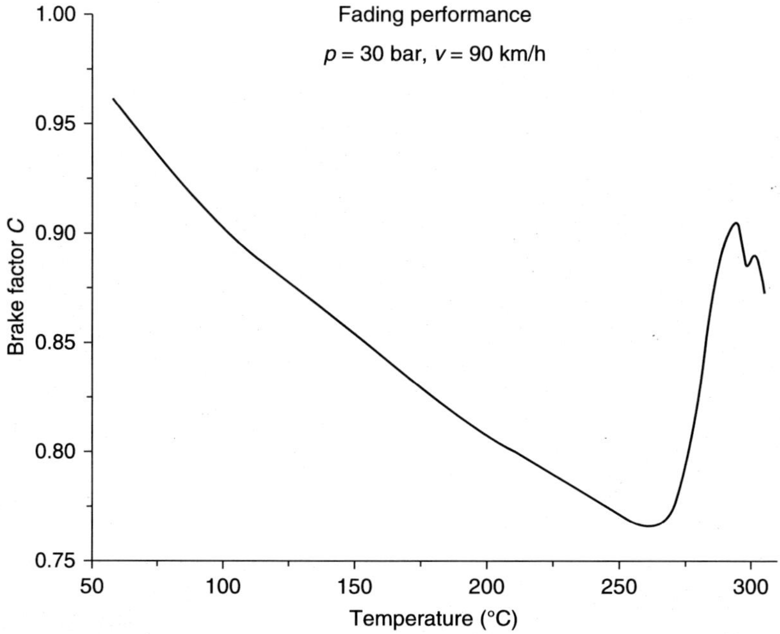

Figure 5.82 **Fading performance of friction material F_{T1} [191]**

behaviour of F_{T1} can be considered as very complex from the point of view of prediction. Firstly, the brake factor decreases from 0.961 to 0.755 in the temperature range between 58 and 272 °C, and after that its value increases to 0.894 at 302 °C. Finally, for the last brake application in the fading test, the brake factor decreases again to 0.873 at 305 °C. The increase in the brake factor above 272 °C is connected with the process of thermal decomposition of the resin finishing. The temperature during each test using the full-scale inertia dynamometer was measured by a thermocouple installed in the inner disc pad 1 mm below the friction surface. These temperature values are related to the average value of the temperature reached during the braking cycle in each brake application.

The influence of the characteristics of the resin on the fading performance of the friction materials can be considered as critical, together with the manufacturing conditions used. The binder resin used for manufacturing the friction materials here had the following properties: flow distance 32 mm, B-time 110 s, hexa content 9% and free phenol content 2%. These resin parameters have a dominant influence on the porosity of the friction material and on its final properties, especially regarding the fading performance. The porosity of a friction material is known to be a crucial factor for heat dissipation and for preventing fading; it is affected mostly by the percentage of binder resin [157]. Furthermore, the fading behaviour can be influenced by the B-time and hexa content because of their influence on the hot press moulding time, degassing behaviour and flow properties. Taking into consideration that 90 neural models were investigated, the quality of prediction of the neural networks was again evaluated by the difference (expressed as a percentage) between the predicted and real values of the brake factor *C*. Six error intervals were chosen (0–5%, 5–10%, 10–15%, 15–20%, 20–25% and 25–30%). Based on the calculated errors between the predicted and real values, the numbers of predicted results which lay in each of these error intervals were evaluated and expressed as a fraction (percentage) of the test data set.

The results for prediction of the fading performance of the friction material F_{T1} can be seen in Figure 5.83. These results are related to the neural models obtained by training (using the training data set derived from F1–F4) the selected neural network architectures using the Bayesian

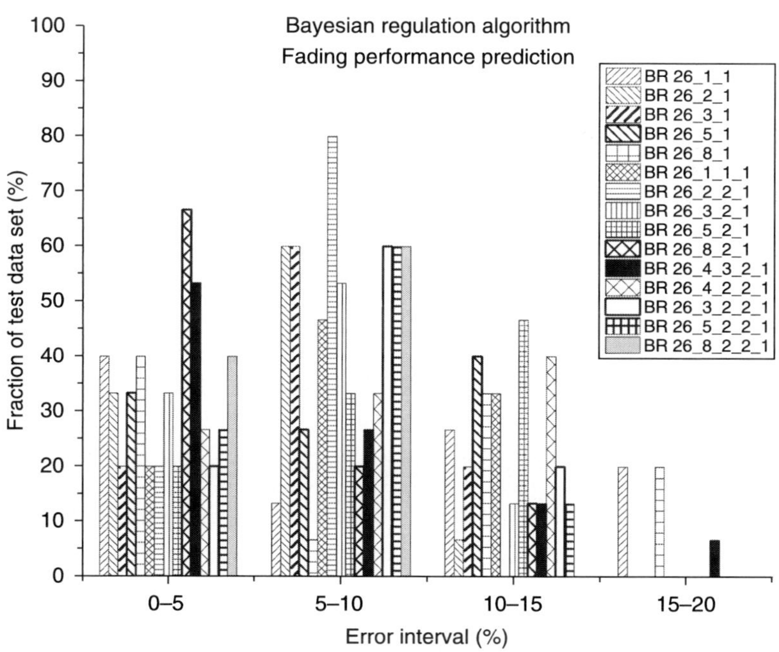

Figure 5.83 **Prediction of fading performance of material F_{T1} by Bayesian regulation algorithm [191]**

regulation algorithm. From Figure 5.83, it is evident that the number of hidden layers and the number of neurons in each of them influence the quality of prediction differently. The one-layered networks show a relatively low percentage of the predicted results in the first error interval. It can be seen that as the number of neurons in the hidden layer is increased, the percentage of predicted results decreases from 40% to 20% (BR 26_3_1) and after that increases again. In the second error interval, the one-layered neural networks with two (BR 26_2_1) and three (BR 26_3_1) neurons in the hidden layer achieve the best result. The two-layered neural networks trained by the Bayesian algorithm show better results when the number of neurons in the first hidden layer is increased. Here, the neural network BR 26_8_ 2_1 achieves the best result, with 66.7% of the predicted results in the first error interval (0–5%) and only 20% of the predicted results in the second error interval. The three-layered neural networks show similar behaviour. When the number of neurons in the first hidden layer is increased from 3 to 8, the percentage of predicted results in the first and second error intervals is increased. From Figure 5.83 it is evident, considering the error interval between 0 and 10%, that the neural network BR 26_8_2_2_1 trained by the Bayesian algorithm shows the best result, with 100% of predicted results with an error less than or equal to 10% (40% between 0 and 5%, and 60% between 5 and 10%).

In order to illustrate the abilities of the neural models BR 26_3_1 and BR 26_8_2_2_1 (based on the training data set F1–F4) to predict the fading performance of the friction material F_{T1}, a comparison between the real and predicted results is shown in Figure 5.84 and Table 5.37. Figure 5.84 shows that these neural models are not able to predict the complex changes in the brake factor C during consecutive braking cycles. Both of the neural models generalize the fact that the brake factor decreases when the brake interface temperature increases. The neural model BR 26_3_1, with only three neurons in the hidden layer, shows a more linear variation of the brake factor. The minimum value of the brake factor predicted by this model is 0.796 at 305 °C, versus the real value of 0.755 at 272 °C (see Table 5.37). Although the neural model BR 26_8_2_2_1 shows more non-linearity than the model BR 26_3_1, the predicted results are similar. Obviously, the influence of the number of hidden layers and the number of neurons in each of them does not provide enough non-linearity to be able to predict the real changes in the brake factor, especially regarding the extreme points. This can be explained by the fact that the output results, for the training data set F1–F4, were linearized in a small range because the prediction abilities of the neural

Table 5.37 Changes in brake factor *C* versus temperature – comparison between real and predicted values [191]

Braking cycle number.	1	2	3	4	5	6	7	8	9	10	11	12	13	14	15
Temperature (°C)	58	97	141	182	216	249	272	283	286	294	295	298	299	302	305
Real	0.961	0.901	0.864	0.821	0.796	0.773	0.755	0.853	0.886	0.908	0.904	0.883	0.886	0.894	0.873
Predicted (BR 26_3_1)	0.922	0.917	0.896	0.874	0.852	0.830	0.813	0.809	0.809	0.804	0.804	0.800	0.800	0.800	0.796
Predicted (BR 26_8_2_2_1)	0.922	0.922	0.891	0.852	0.835	0.822	0.817	0.817	0.817	0.817	0.817	0.817	0.817	0.817	0.813

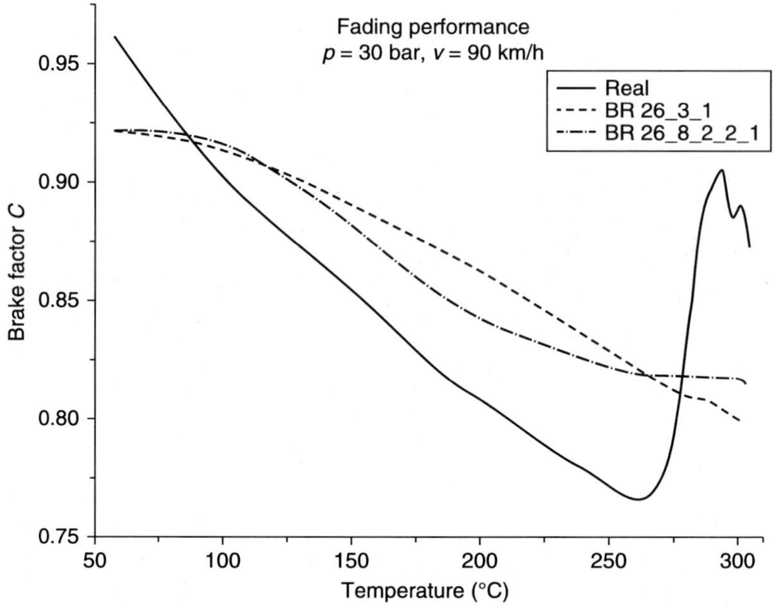

Figure 5.84 **Prediction of fading performance of material F_{T1} – real and predicted by Bayesian regulation algorithm [191]**

models were limited to a linear or nearly linear approximation. As mentioned earlier, such a linearization was introduced because of the different nature of the data that was stored in the training data set (cold, fading and recovery performance). Furthermore, out of 360 training data points, only 60 data points were related to the fading performance.

Figure 5.85 shows the predictions of the fading performance using the same neural networks but now trained using the resilient backpropagation algorithm. This new training algorithm, for the same networks and training data set, influenced the prediction capabilities of the neural networks differently. It is evident from Figure 5.85 that a higher percentage of the predicted results lie in the first error interval, especially in the case of the predictions of the neural models obtained from three-layered neural networks. The one-layered neural networks are very sensitive to changes in the number of neurons in the hidden layer (Figure 5.85). In the case of the two-layered neural networks, the percentage of predicted results in the first error interval increases when the number of neurons in the hidden layer is by increased. The three-layered networks trained by the resilient backpropagation algorithm

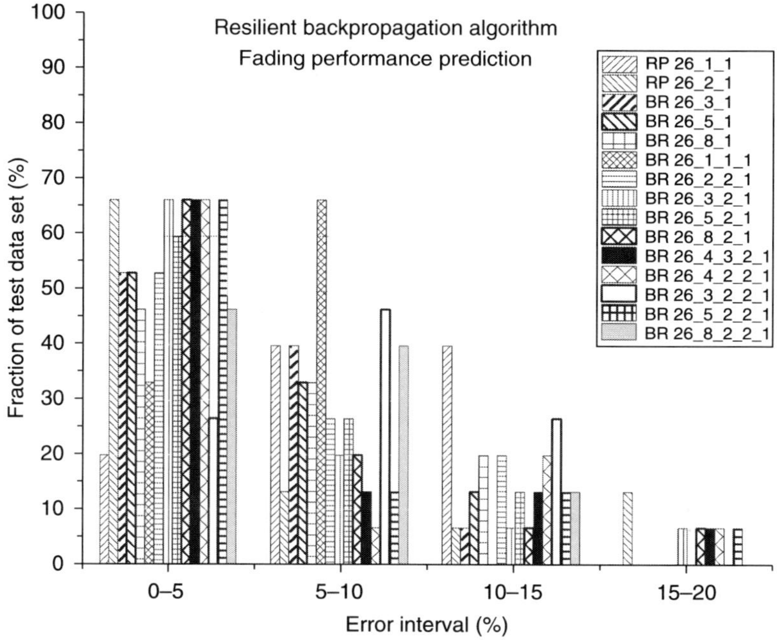

Figure 5.85 **Prediction of fading performance of material F_{T1} by resilient backpropagation algorithm [191]**

show the best stability with respect to changes in the number of neurons. Although training with the resilient backpropagation algorithm nominally shows better results regarding the number of predicted results in the first error interval, the fading performance is not generalized well by the neural networks (only on an average level). Similar prediction results were achieved by the other training algorithms (Levenberg–Marquardt, scaled conjugate gradient and gradient descent). After testing the neural networks with the first test data set, obtained from F_{T1}, the same neural networks were tested with the second test data set, obtained from F_{T2}. Again, the best results were achieved by the three-layered neural network (26 8 2 2 1) trained by the Bayesian regulation algorithm. The results for prediction of the fading performance of the friction material F_{T2} are shown in Figure 5.86. The neural model BR 26_8_2_2_1 shows more non-linearity, but this model is not able to predict the real changes in the brake factor C versus the brake interface temperature.

When changes were introduced in the training data set, the sensitivity of the neural networks tested was improved. The quality of prediction was evaluated regarding the following criteria: (i) prediction of the value of the brake factor C during the first brake application, (ii) prediction of

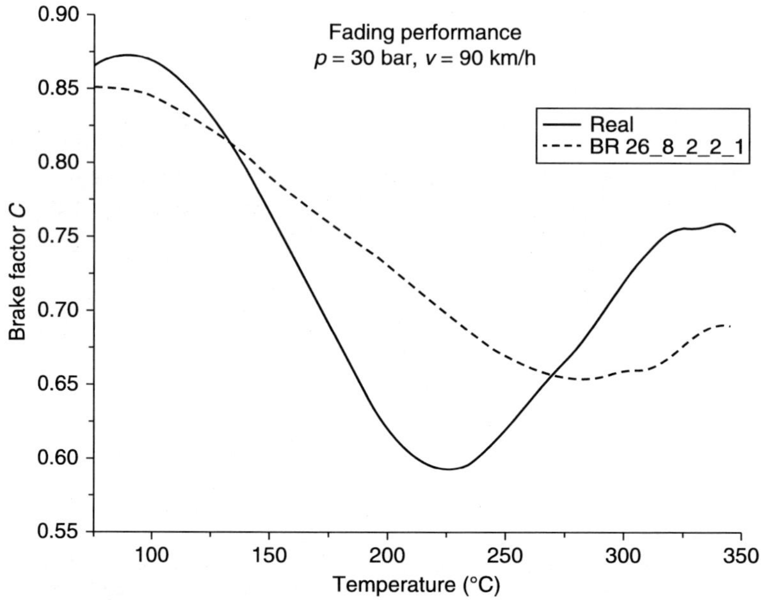

Figure 5.86 **Prediction of fading performance of material F_{T2} – real and predicted by Bayesian regulation algorithm [191]**

the minimum value of the brake factor and the total amount of fading, (iii) prediction of the temperature for the minimum value of the brake factor, (iv) prediction of the maximum value of the brake factor after it had reached its minimum, and the temperature at which that maximum was reached, and (v) prediction of the brake factor during the last brake application.

Based on these criteria, after training and testing of 90 neural models, the three neural models BR 26_8_4_2_1, BR 26_8_4_1 and BR 26_10_5_1, trained by the Bayesian regulation algorithm, showed the best results. The predictions of the fading performance of friction material F_{T1} by the neural model BR 26_8_4_2_1 are shown in Figure 5.87. This neural model was obtained using a three-layered neural network with eight neurons in the first hidden layer and four and two neurons in the second and third hidden layers, respectively. Figure 5.87 shows that the sensitivity of the neural network, i.e. the non-linearity of the predicted results, was significantly improved. As shown in Figure 5.87 and Table 5.38, the quality of the prediction of fading performance obtained by means of the neural model BR 26_8_4_2_1 was evaluated. It is evident that this neural model generalizes well the behaviour of the friction

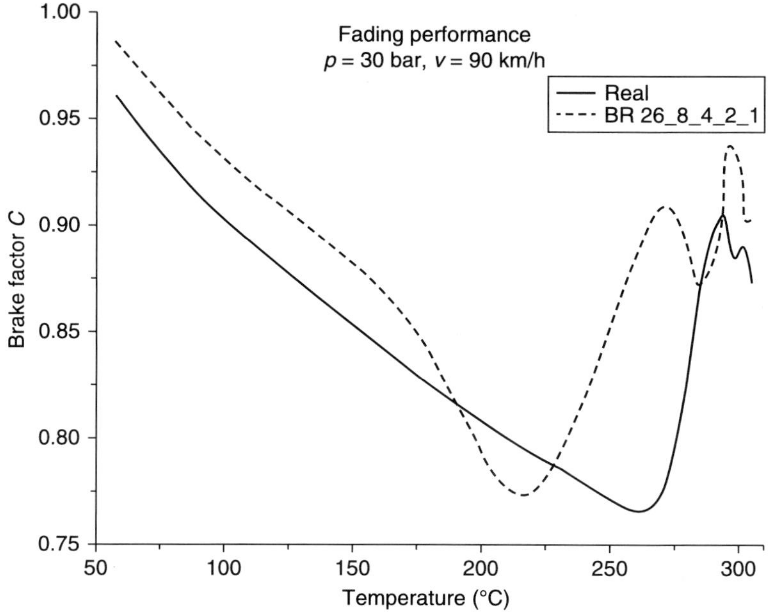

Figure 5.87 **Prediction of fading performance of material F_{T1} – real and predicted by neural model BR 26_8_4_2_1 (training data set F1–F8) [191]**

material F_{T1} at elevated temperatures. The initial value of the brake factor *C* and its minimum value during the fading test are well predicted (see Table 5.38). It can be seen that the difference between the real and predicted values of the fading is relatively small (a real value of 21.4% versus a predicted value of 25.4%). The fading value was calculated as the difference between the maximum and minimum values of the brake factor during the fading test. The predicted value of the temperature which corresponds to the minimum value of the brake factor is 216 °C, compared with the real value of 272 °C. Obviously, the minimum value of the brake factor is predicted at 56 °C earlier than for the real friction material. Furthermore, the tendency of the friction material F_{T1} to increase its performance after reaching the minimum value of the brake factor is also well predicted. There is a difference between the actual and predicted values of the maximum brake factor after the minimum value is reached, as well as between the temperatures of the maximum. The real value of the brake factor increases to 0.904 at 294 °C, versus the predicted value of 0.933 at 272 °C.

Table 5.38 Comparison of the prediction capabilities of neural models [191]

Neural model	Criteria				
	Brake factor in 1st brake application	Minimum value of brake factor/fading value (%)	Temperature for minimum brake factor (°C)	Maximum value of brake factor after fading/temperature (°C)	Brake factor in last brake application
Real	0.961	0.755/21.4	272	0.904/294	0.873
BR 26_8_4_2_1	0.986	0.737/25.2	216	0.933/272	0.904
BR 26_8_4_1	0.915	0.8/12.6	182	0.902/294	0.926
BR 26_10_5_1	0.92	0.765/19	249	0.859/295	0.855

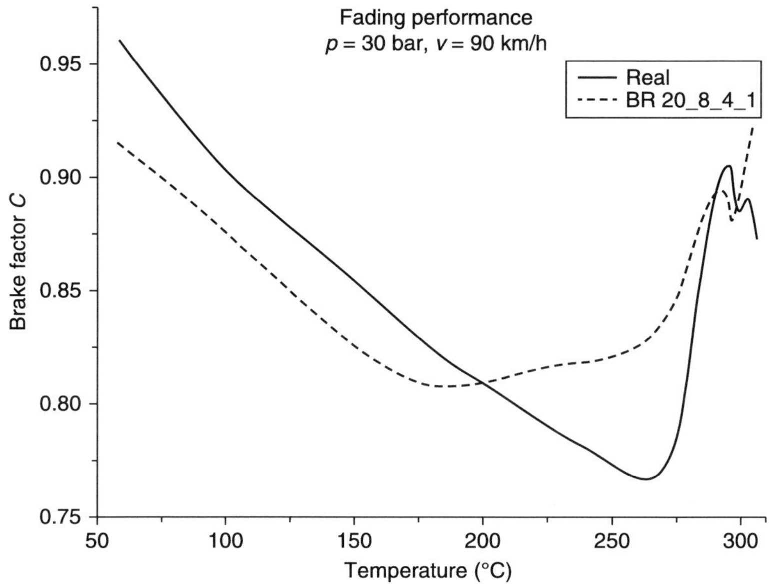

Figure 5.88 **Prediction of fading performance of material F_{T1} – real and predicted by neural model BR 26_8_4_1 (training data set F1–F8) [191]**

The prediction performance of the neural model BR 26_8_4_1 obtained by training a two-layered neural network is shown in Figure 5.88. Using the criteria established for the evaluation of the prediction capabilities of neural models, it can be concluded that the predictions of the fading performance for the same friction material (F_{T1}) are influenced differently by the architecture of the new neural network. The initial and minimum values of the brake factor are not well predicted; indeed, the predicted value of the fading is 12.6%, versus the real value of 21.4%. Furthermore, the minimum value of the brake factor occurs at 182 °C, versus the real value of 272 °C. The recovery of the brake factor after it has reached its minimum value is predicted well, with a predicted value of 0.902 at 294 °C, versus the real value of 0.904 at 294 °C.

Regarding the neural models for the prediction of fading performance described previously, it can be seen that the neural model BR 26_8_2_1, based on a three-layered neural network, has more propensity towards overfitting than the neural model based on a two-layered neural network BR 26_8_4_1. On the other hand, the capability of the neural model BR 26_8_4_1 is not enough to approximate the complex changes in the brake factor *C* versus temperature. That is why the numbers of neurons

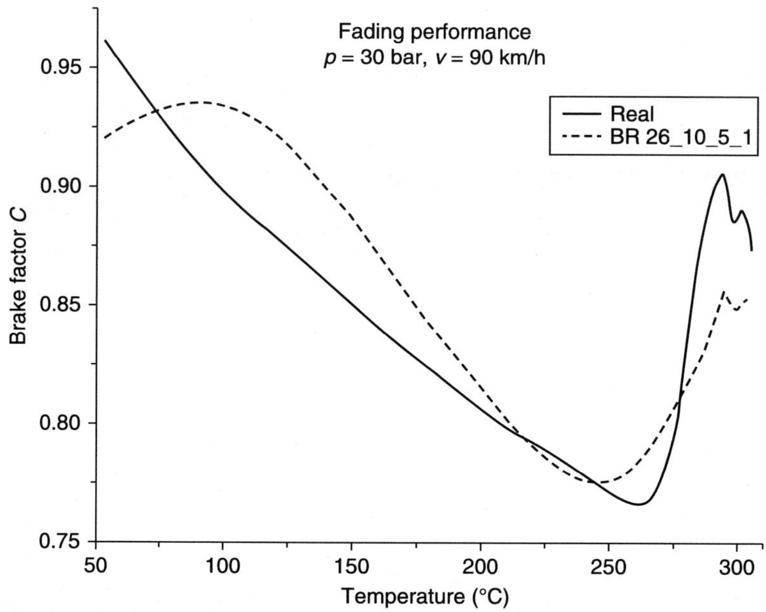

Figure 5.89 **Prediction of fading performance of material F_{T1} – real and predicted by neural model BR 26_10_5_1 (training data set F1–F8) [191]**

in the hidden layers of the two-layered neural network were increased. The results of the prediction of fading performance are shown in Figure 5.89. Taking into consideration the criteria for evaluation of the prediction capabilities of neural models (see Table 5.38), it can be seen that the neural model BR 26_10_5_1 shows the best prediction with respect to all criteria in Table 5.38. The predicted value of the fading is 19%, versus the real value of 21.4%. The minimum value of the brake factor *C* is also well predicted, as 0.765 at 249 °C versus the real value of 0.755 at 272 °C. The recovery of the brake factor after it decreases is also predicted at an acceptable level, with a predicted value of 0.859, versus the real value of 0.904 at the same temperature.

The performance of the best neural model, BR 26_10_5_1, was tested using the second test data set, i.e. that for F_{T2}. The fading performance of the friction material F_{T2}, shown in Figure 5.90, is significantly different from that of the friction material F_{T1}. The neural model BR 26_10_5_1 generalizes well the behaviour of the new friction material F_{T2} at elevated temperatures. In Table 5.39, the real and predicted results are compared. As evidenced, all of the criteria established are satisfied.

Table 5.39 Comparison of the prediction capabilities of neural model BR 26_10_5_1 versus real data [191]

Neural model	Criteria				
	Brake factor in 1st brake application	Minimum value of brake factor/fading value (%)	Temperature for minimum brake factor (°C)	Maximum value of brake factor after fading/temperature (°C)	Brake factor in last brake application
Real	0.867	0.557/36.7	230	0.757/320	0.754
BR 26_10_5_1	0.84	0.6/31	236	0.77/329	0.773

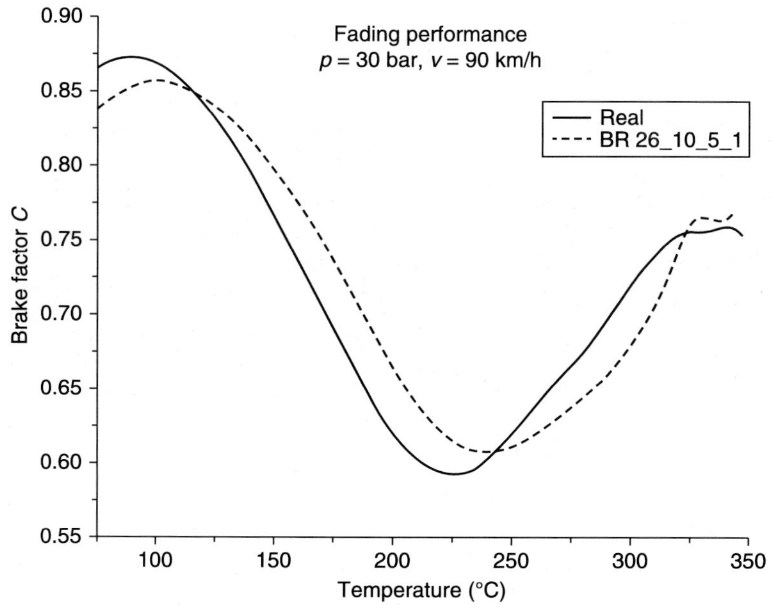

Figure 5.90 **Prediction of fading performance of material F_{T2} – real and predicted by neural model BR 26_10_5_1 (training data set F1–F8) [191]**

5.3.3 *Modelling and prediction of brake friction material wear*

Introduction

The phenomenon of material transfer during sliding is important from both a scientific and a practical point of view. Wear is not only a particle detachment mechanism; wear performance is influenced by how wear particles act inside the contact zone as a third body and how they are ejected from the contact zone [194]. Friction and wear performance are two kinds of response of a tribo-system [195]. Wear in general depends on many factors, such as the temperature, the applied load, the sliding velocity, the properties of the mating materials and the durability of the transfer layer. The transfer layer affects the friction and wear behaviour of the friction pair [196]. Remarkable effort has been put into developing appropriate wear models and predicting the wear properties of engineering materials. According to [197], the wear volume can be predicted as a function of the normal load, the sliding velocity and the hardness of the material by the equation $V = KPL/H$, where V is the volume of worn material (in m^3), K is the wear coefficient (to be determined experimentally), P is the applied load (in N), L is the sliding distance (in m) and H is the hardness of the material. Prediction of the wear volume of a brake lining material was investigated in [198]. In the model used, the wear volume was taken as proportional to power functions of the load, the sliding velocity and the sliding time, so that the wear volume was given by $V = K_f P^a V^b t^c$, where V is the wear volume (in m^3), K_f is the wear coefficient of the friction material (to be determined experimentally), P is the applied load (in N), V is the sliding velocity (in m/s), t is the sliding time (in s), and a, b and c are exponents that depend on the material and the sliding conditions [198]. Thus, in classical theories, the wear volume V is proportional to the normal load N, the sliding distance L and the inverse of the hardness H [197], and the wear coefficient K can be used as a measure of the wear behaviour of a material. Both K and H reveal the response of the material and thus can be grouped into a single parameter called the specific wear rate, $k = V/SN$, where S is the sliding distance [199]. The equation defined in [198] was modified to $h = kP^a(\Omega r)^b t^c$, where h is the wear displacement, P is the normal contact pressure, Ω is the rotational speed of the disc (in rad/s), r is the mean radius of the pad (m), and a, b and c are all constants which remain to be determined [200].

The wear mechanisms operating in brake lining materials have been found to be a complex combination of abrasion, adhesion, fatigue, delamination and thermal decomposition [201]. These wear mechanisms and their transitions from one combination to another determine the stochastic nature of the wear of brake friction materials. The dominant wear mechanism of brake friction materials is influenced by the braking time, the applied load and material characteristics. Accordingly, the prediction of wear performance, especially that of brake friction materials, is a very complex task. The presence of a transfer film generated in the contact area of the friction pair further complicates the situation. The transfer film in polymer–metal sliding contacts is generally believed to be responsible for the gradual transition from transient to steady-state wear behaviour. However, the role of the sliding variables such as the velocity, load and temperature in the transient wear behaviour, as well as their effect on the transfer film, is not clear. Thus the application of polymer composites as engineering materials has become state-of-the-art [201–203]. The investigation of their properties and the achievement of a satisfactory understanding of their behaviour play a key role in the design of composites for special engineering applications. The simulation of material properties generally involves the development of a model derived from experimental data [181, 204–208]. The lack of good prediction methods is related to objective difficulties in the mining of experimental data to describe and understand the complex behaviour of brake friction materials and capture all of the relevant influences on their wear properties. The traditional mathematical techniques that have been used to model the wear of brake friction materials in previous studies [198–200] are not able to provide a wear model with an inherent ability to generalize the complex synergy of different influences of the formulation and manufacturing conditions of brake friction materials, and of the load conditions on the brake. According to [193], artificial neural networks are suited for simulating complex polymer composite problems because they can be trained to find solutions [208, 209]. In this section, we describe the use of an artificial neural network approach for prediction of the wear of brake friction materials, taking into account all relevant influencing factors, where we manufactured the materials and performed wear testing with a full-scale inertia dynamometer. The wear was predicted by developing a neural model of the wear of friction materials. This neural model was used to provide a functional approximation between inputs consisting of the formulation of the friction material, the manufacturing conditions and the operation conditions, and an output consisting of the changes in the wear volume of the material.

Wear testing

The development of the neural model of the wear of friction materials described here was based on experimental data; the experimental phase had to be planned carefully in order to provide input/output values for training, validation and testing of the neural networks. The wear-testing methodology needed to be chosen according to the range and distribution of data that would be collected (Table 5.40), and had to be properly designed in order to provide the average brake operation conditions. As can be seen in Table 5.40, the testing methodology was designed in such a way that the brake actuation pressure and initial speed were constant for all wear tests (initial speed, 90 km/h; brake actuation pressure, 40 bar). This would provide relatively average braking regimes in all wear tests (in terms of the work done in a brake application) for different brake interface temperatures. The brake interface temperature was varied from 100 °C (wear test 1) to 250 °C (wear test 3). A gradual transition from a transient to a steady-state wear condition of the friction material was produced before the wear test (Table 5.40). This was done by performing the following tests: (i) burnishing (150 brake applications),

Table 5.40 Wear-testing methodology [208]

Test conditions	Applied pressure (bar)	Initial speed (km/h)	Initial brake interface temperature (°C)	Number of brake applications
Initial burnishing	40	90	<100	150
Wear test 1	40	90	≤100	70
Wear test 2	40	90	≤175	50
Wear test 3	40	90	≤250	30

(ii) a cold-performance test (25 brake applications; $p = 20$–100 bar and $v = 20$–100 km/h), and (iii) a fade and recovery test (40 brake applications). The number of brake applications in each wear test was chosen in order to obtain a measurable amount of wear in a short period (Table 5.40).

Input data

A neural model of the wear of friction materials should be able to capture three groups of input data: (i) the formulation of the friction material, (ii) the manufacturing conditions and (iii) the working regime of the brake. The total number of input parameters in the artificial neural networks was 25. The influence of the formulation was represented by 18 input parameters related to the volume percentage of each ingredient in the formulation (Table 5.41). The influence of the manufacturing conditions was represented by five parameters related to the most important manufacturing factors (Table 5.42), the values of which were scaled between 0 and 1. The operation conditions of the friction material were represented by two parameters, related to the brake interface temperature and the work done in a brake application. Thus, changes in the measured parameters during braking, such as the brake actuation pressure, the speed and the braking torque, determined the work done in a brake application. This input parameter was chosen to represent the operation conditions of the brake in synergy with the brake interface temperature, which was also measured. After the identification of the input parameters and their number, the next important decision was related to the ranges of the input parameters. The range and distribution of the data related to the working regime of the brake were defined by the testing methodology (Table 5.40). On the other hand, the ranges and distributions of the manufacturing and, especially, the formulation

Table 5.41 Ranges of proportions of raw materials used (vol%) [208]

Ingredient	Friction materials F1–F7 (training set)	Test data sets		
		F_{T1}	F_{T2}	F_{T3}
Phenolic resin	17–25	25	17	20
Iron oxide	3–7	5	3	5
Baryte	26–15	15	26	25
Calcium carbonate	1–3	3	1	2
Brass chips	1–3	3	1	2
Aramid	2–6	6	2	1
Mineral fibre	10–16	10	9	14
Vermiculite	4–8	8	4	6
Steel fibre	4–1	1	4	2
Glass fibre	2–4	4	2	2
Brass powder	1–2	2	1	2
Copper powder	1–3	3	1	2
Graphite	8–3	3	7	4
Friction dust	5–2	2	5	5
Molybdenum disulphide	1–3	3	1	1
Aluminium oxide	2–3	3	2	2
Silica	1–2	2	1	1
Magnesium oxide	8–2	2	8	4

parameters of the friction materials presented a more difficult task. The input data related to the formulation of the friction materials and their manufacturing conditions were selected as listed in Tables 5.27 and 5.28. The data shown in Tables 5.41 and 5.42 were used as the neural-network inputs. These data had to be chosen depending on the types of friction material and the ranges of proportions of ingredients used. The manufacturing parameters for which the neural model was going to be developed were chosen similarly.

The formulation and manufacturing parameters for each type of friction material, as shown in Tables 5.41 and 5.42, were completely different from one another. The input data obtained from the manufacture of the friction materials denoted by F1–F7 were used for training of the

Table 5.42 **Manufacturing conditions [208]**

Manufacturing parameter	F1–F7 (training set)	Test data set		
		F_{T1}	F_{T2}	F_{T3}
Moulding pressure (MPa)	45–70	40	70	55
Moulding temperature (°C)	155–175	170	155	155
Moulding time (min)	6–11	11	6	8
Heat treatment temperature (°C)	200–250	200	200	220
Heat treatment time (h)	12–5	12	5	7

neural networks, while the input data obtained from the manufacture of the materials F_{T1}, F_{T2} and F_{T3} were used for testing the prediction capabilities of the networks. The data for testing the neural networks were obtained by manufacturing three new types of friction materials whose input parameters were completely different from those stored in the training data set. The formulation and manufacturing parameters of the friction materials used for testing were chosen to be at the end of, in the middle of or outside the specified range (see Tables 5.41 and 5.42). This was done in order to test the neural model over the whole range of input parameters. Owing to the unknown interrelated influences of the ingredients in the formulation of the friction material, the manufacturing conditions and the working regime of the brake, a random distribution over the selected ranges was used. Thus, the correlation between the operation conditions and the amount of wear of the friction material was investigated. The work done in a brake application was calculated by measuring the braking torque and speed during a braking cycle (Figures 5.91(a) and (b)). Based on the measured values of the braking torque and speed, the braking power could be determined for each brake application. The work done in a brake application was calculated by integrating the braking power over the braking time (see Figure 5.91). The brake interface temperature, which served as a representation of the operation conditions of the friction material (Figure 5.91(d)), was also measured and used as an input parameter. The initial brake interface temperature was set as described in Table 5.40 before each brake application, according to the particular wear test performed. The synergistic influence of the 25 input parameters on the wear volume is illustrated in Figure 5.92.

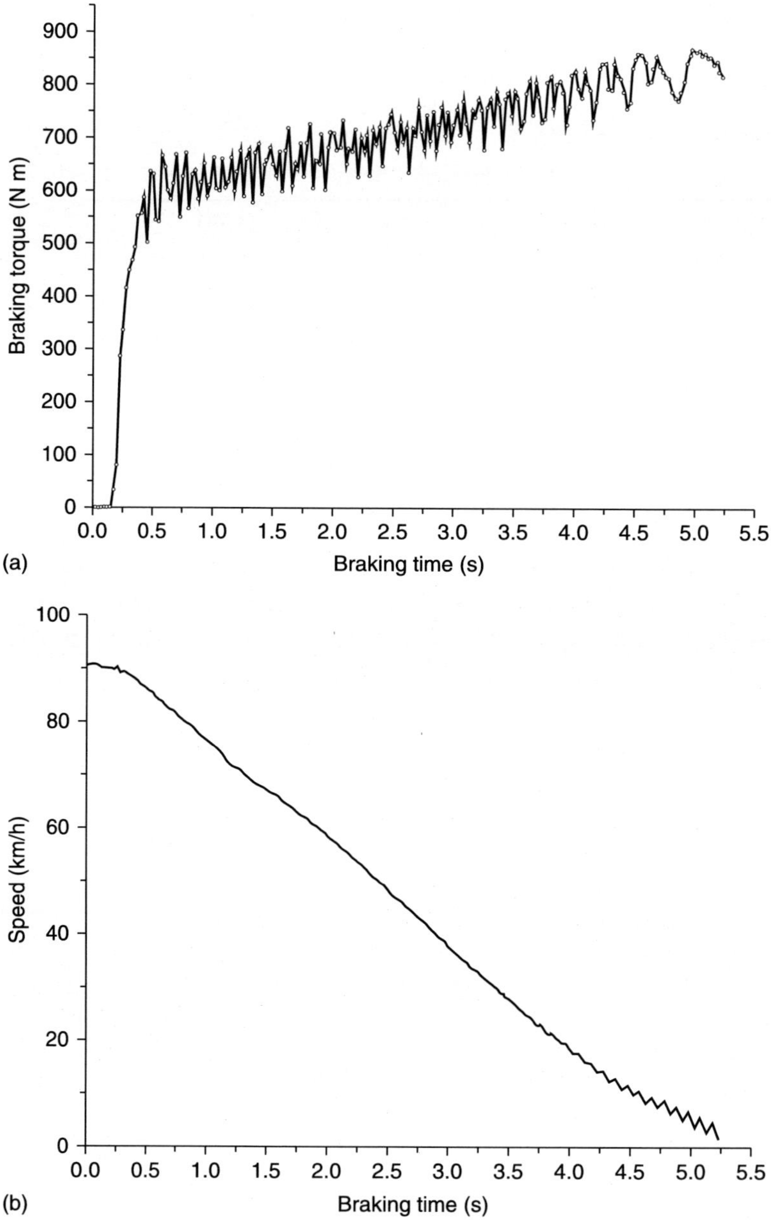
Braking torque (N m)
Braking time (s)
(a)
Speed (km/h)
Braking time (s)
(b)

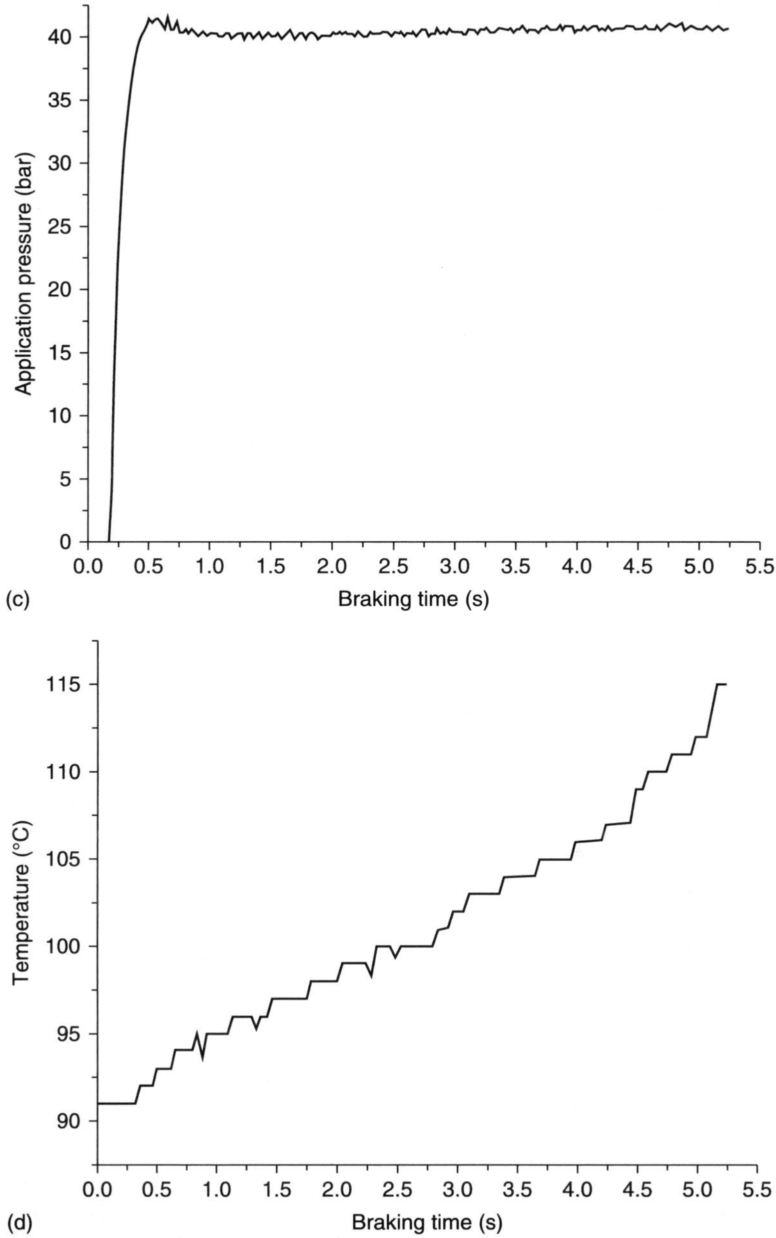

Figure 5.91 **Examples of changes in measured parameters during a braking cycle: (a) braking torque, (b) speed, (c) application pressure and (d) brake interface temperature [208]**

Output data

As illustrated in Figure 5.92, the synergistic influence of 25 input parameters was correlated with the wear volume of the friction materials. In order to be tested, all types of friction material were produced as a disc pad assembly mounted on the front brake of a passenger car, and tested using a single-ended full-scale inertia dynamometer. The wear volume of friction material in the three different wear tests was considered as an output parameter (see Figure 5.92 again); this was calculated after each wear test based on the linear wear of the friction material. The linear wear was measured at the end of each wear test. The wear volume was calculated by taking the average value of the linear wear of the material (in mm), measured at five previously determined points (Figure 5.93)

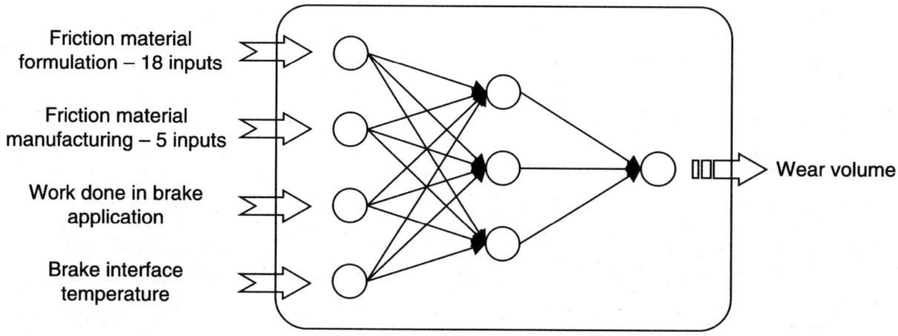

Figure 5.92 Neural modelling of the wear of friction materials [208]

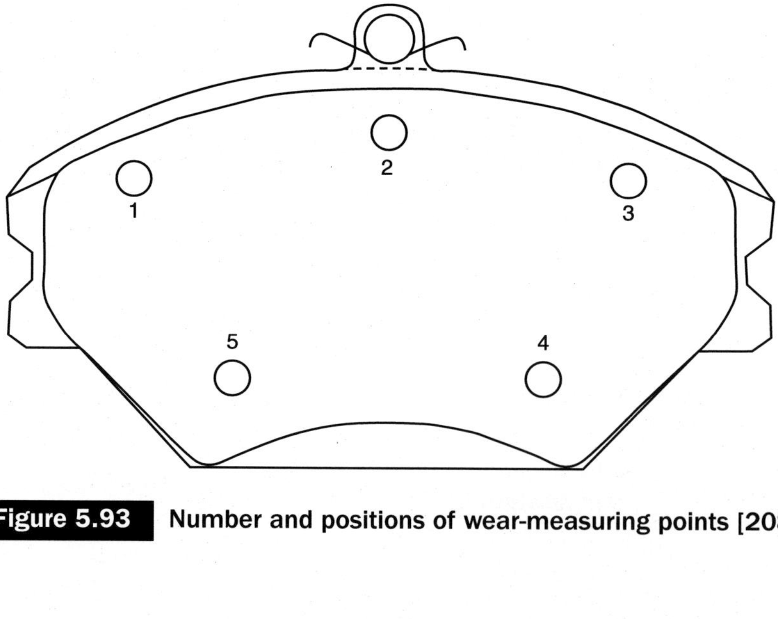

Figure 5.93 Number and positions of wear-measuring points [208]

after each wear test, and multiplying that value by the known nominal friction surface area.

Wear modelling

Neural modelling of the wear of friction materials needs an appropriate architecture of the neural network, as well as a learning algorithm, in order to represent the functional relationship between the inputs and output of the network. The prediction capabilities of a neural network are affected by the representational power of the network and the learning algorithm selected. The learning ability of neural networks, in order to extend their predictive power to data outside the training data set, is essential for the implementation of artificial neural networks. In order for a neural network to learn about the wear properties of a friction material, an appropriate number of input/output training pairs have to be provided. Therefore, sufficient input/output pairs have to be stored in the training data set. The data points for training and testing the neural networks were obtained from the formulation, manufacturing and testing of 10 different types of friction material.

Thus, the number of brake applications in each wear test was prescribed (Table 5.40); with each new brake application, the input parameters to the neural network changed. This was done because the cumulative work done in the brake applications changed with each brake application. Consequently, the linear wear, and therefore the volume wear, of the brake friction material also changed. On the other hand, the linear wear of the friction material became measurable only after a number of brake applications. Thus, since the moment when the linear wear would become measurable was not known in advance, the linear wear of the material was measured after all of the brake applications prescribed in the wear test (as in Table 5.40). Since the synergistic effects of the input parameters on the wear performance of the friction material changed with each new brake application, the number of input values to the neural network corresponded to the number of brake applications in each wear test (70, 50 and 30 input values; see Table 5.40). In contrast, the linear wear of the friction material, the output parameter, was measured after each wear test. This means that one value of the linear wear was provided at the end of a wear test. Three values of the output parameter, i.e. the linear wear (or volume wear) could be measured in the three wear tests, whereas 150 inputs were changed per wear test. Obviously, the number of output results is not enough for a successful training of the neural networks versus the number of input parameters. In order to provide enough wear results, i.e. output values for training of the artificial neural networks, and at the same to shorten the period of testing the friction materials, a linear functional relationship was assumed between the work done in a brake application and the number of braking cycles. The same relationship was assumed between the number of brake applications and the wear of the friction material per brake application [210]. This meant that the wear w obtained after one (the first) brake application became $2w$ after the second brake application and $70w$ after 70 brake applications in wear test 1, for example. A similar approach was used in the case of the other wear tests.

Since the value of the wear of the friction material was known after each wear test, as well as the total work done in all brake applications, the specific wear rate could be determined. The value of the specific wear rate could be obtained by dividing the total volume wear of the friction material by the total amount of work done in all braking cycles (expressed in mm^3/kJ). Based on the specific wear rate determined in this way, the wear volume of the friction material could be calculated after certain amount of work had been done by the brake. This was used to increase the number of output results for each wear test. Based on the specific wear rate of the friction material, the wear volume of the material could be determined for each brake application. This meant that 70 output results (wear volume) could be calculated for wear test 1 (and 50 for wear test 2 and 30 for wear test 3). The 'sampling' of input/output pairs for training of the neural networks was done by choosing values of the wear volume close to the final value measured at the end of the wear test. In order to develop a neural model of the wear of friction materials able to predict the wear volume under conditions that corresponded to steady-state wear, the last 20 output results (between the 50th and 70th braking cycles) from wear test 1, the last 17 results from wear test 2 (between the 33rd and 50th braking cycles) and the last 10 results from wear test 3 (between the 20th and 30th braking cycles) were chosen.

This 'methodology' of sampling the wear data of the brake friction materials provided 47 wear data points per friction material in all of the wear tests (20+17+10=47). Since seven different types of friction materials were available for training the neural networks, 329 input/output data points were used for training. So that the neural networks could learn easily about the wear performance of the friction materials in different braking regimes, the training data set was organized in such a way that it caused minimal disturbance to the training of the neural networks. The training data set was composed of seven groups of input/output data, where each group was related to a specific type of friction material, denoted by F1–F7. The first group of input/output data was related to 47 data points for the friction material F1 (20 data points related to wear test 1, followed by 17 data points related to wear test 2, followed by 10 data points related to wear test 3). This group of data was accompanied by 47 data points related to the friction material F2, followed by data related to material F3, up to the data related to the friction material F7 at the end of the training data set. Organized in this way, the training data set was used for training different architectures of neural networks. The capabilities of the neural networks to predict and generalize the wear behaviour of the friction materials were tested by the test data set. The test data set was created using input/output data obtained from the formulation, manufacturing and testing of the three new types of friction material denoted by F_{T1}, F_{T2} and F_{T3} (Tables 5.41 and 5.42). Since 47 wear data points were chosen for each combination of formulation, manufacturing and testing of a friction material, 141 input/output data points were provided for testing the neural networks (with the three new types of friction material). The wear properties the friction materials F_{T1}, F_{T2} and F_{T3} in wear tests 1, 2 and 3 are shown in Table 5.43. The specific wear rate was calculated taking into consideration the total work done in the brake applications in the wear test. The total work in the wear test was determined by summing all of the average values of the work done by the brake in each wear test. The volume wear was determined using the value of the linear wear measured after each wear test.

Again, a trial and error method was employed to find the best neural-network characteristics for matching the particular input/output relationship. Based on MATLAB 6.5, Release 13, the following neural-network architectures were investigated: (i) the one-layered architectures 25 $[1]_1$ 1, 25 $[2]_1$ 1, 25 $[3]_1$ 1, 25 $[5]_1$ 1 and 25 $[8]_1$ 1; (ii) the two-layered architectures 25 $[1\text{-}1]_2$ 1, 25 $[2\text{-}2]_2$ 1, 25 $[3\text{-}2]_2$ 1, 25 $[5\text{-}2]_2$ 1 and 25 $[8\text{-}2]_2$ 1; and (iii) the three-layered architectures 25 $[3\text{-}2\text{-}2]_3$ 1, 25 $[4\text{-}3\text{-}2]_3$ 1, 25 $[4\text{-}2\text{-}2]_3$ 1, 25 $[5\text{-}2\text{-}2]_3$ 1 and 25 $[8\text{-}2\text{-}2]_3$ 1. These 15 neural-network architectures were trained by the Levenberg–Marquardt, Bayesian regulation, resilient backpropagation, scaled conjugate gradient and gradient descent training algorithms. The same sigmoid activation function as before (Eq. 5.86) was used between the input and hidden layers. A linear activation function was employed between the hidden and output layers. Pre-processing of the input parameters was carried out before the training of the neural networks. The 18 parameters related to the formulation of the friction material were presented to the network as volume percentages. The input data related to the five manufacturing

Table 5.43 Wear results for the friction materials F_{T1}, F_{T2} and F_{T3} [208]

Wear properties	Wear test		
	Test 1	Test 2	Test 3
Friction material F_{T1}			
Total volume wear (mm^3)	376.74	566.72	660.1
Average value of work done in a brake application in the wear test (J)	134562	137843	140180
Total work done in the wear test (kJ)	5247.9	4548.8	4205.4
Specific wear rate (mm^3/kJ)	0.072	0.125	0.157
Friction material F_{T2}			
Total volume wear (mm^3)	2975.28	2067.24	3442.18
Average value of work done in a brake application in the wear test (J)	130852	130273	130439
Total work done in the wear test (kJ)	9159.7	6513.684	3913.174
Specific wear rate (mm^3/kJ)	0.325	0.317	0.880
Friction material F_{T3}			
Total volume wear (mm^3)	2350.6	2002.84	1513.4
Average value of work done in a brake application in the wear test (J)	127100	127433	126809
Total work done in the wear test (kJ)	8897.051	6371.679	3804.279
Specific wear rate (mm^3/kJ)	0.264	0.314	0.398

parameters and to the brake operating regime (the work done by the brake and the brake interface temperature) were scaled in the range from 0 to 1 according to Eq. 5.87. The output parameter (wear volume) was linearized by the following equation:

$$O_{\mathrm{Lin}} = 0.8 + 0.01 \times \frac{(O_{\mathrm{Curr}} - O_{\mathrm{Min}})}{(O_{\mathrm{Max}} - O_{\mathrm{Min}})}, \qquad [5.91]$$

where O_{Curr} is the current output value, O_{Max} is the maximum output value and O_{Min} is the minimum output value.

Prediction results

The trained neural networks were employed to predict the wear of the brake friction materials F_{T1}, F_{T2} and F_{T3}. Fifteen different neural networks were trained using the five aforementioned training algorithms. As a consequence, 75 neural models were developed for prediction of the wear of friction materials. As mentioned earlier, all of these neural models were tested with unknown input/output parameters obtained from the formulation, manufacturing and testing of the friction materials F_{T1}, F_{T2} and F_{T3}. The quality of prediction of the wear behaviour of friction materials was evaluated taking the following criteria into consideration: (i) the quality of prediction of the wear volume of the friction material versus changes in the brake interface temperature at the end of each wear test (the specific wear rate), and (ii) the quality of prediction of the wear volume versus changes in the work done in brake applications during the wear test. The best results for prediction of the wear behaviour of friction materials were shown by the neural model denoted by BR 25_5_1. This neural model was obtained by training of the one-layered neural network 25 $[5]_1$ 1 (25 inputs, five neurons in the hidden layer and one output) with the Bayesian regulation algorithm.

The neural model was first tested considering the formulation and manufacturing input parameters related to the friction material F_{T1}. The total work done in the brake applications was 5247.9 kJ (Table 5.43) for an initial brake interface temperature $T \leq 100\,°C$. The same neural model was also tested for the formulation and manufacturing parameters of the friction material F_{T1} and the following inputs: (i) the total work done in brake applications was equal to 4548.8 kJ, $T \leq 175\,°C$ (wear test 2), and (ii) the total work done in brake applications was equal to 4205.4 kJ, $T \leq 250\,°C$ (wear test 3). Figure 5.94 shows a comparison between the real and predicted specific wear rates. Similarly, the neural model was used for prediction of the specific wear rates of friction materials F_{T2} and F_{T3}, with inputs related to the work done by the brake and to the brake interface temperature as specified in Tables 5.40 and 5.43. As shown in Figure 5.94, there are significant differences between the real wear volumes, i.e. the real specific wear rates, between the three friction materials tested (F_{T1}, F_{T2} and F_{T3}). This is especially evident at elevated temperatures (Figure 5.95, $T = 250\,°C$), where the friction material F_{T2} shows a high value of the specific wear rate. This is caused by complex influences of the formulation and manufacturing conditions of material F_{T2} in synergy with the braking regime. From Tables 5.41 and 5.42, it can be seen that the formulation of the friction material F_{T2} had a lower

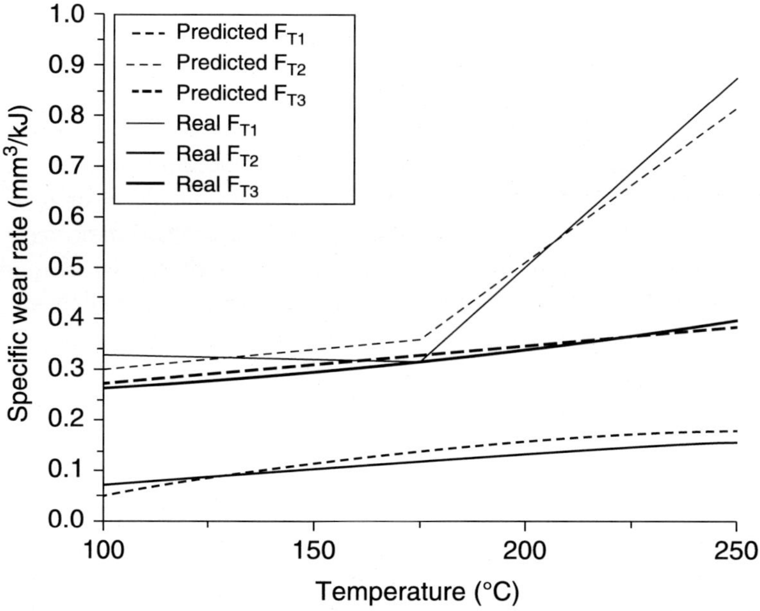

Figure 5.94 Comparison between real and predicted specific wear rates (friction materials F_{T1}, F_{T2} and F_{T3}) [208]

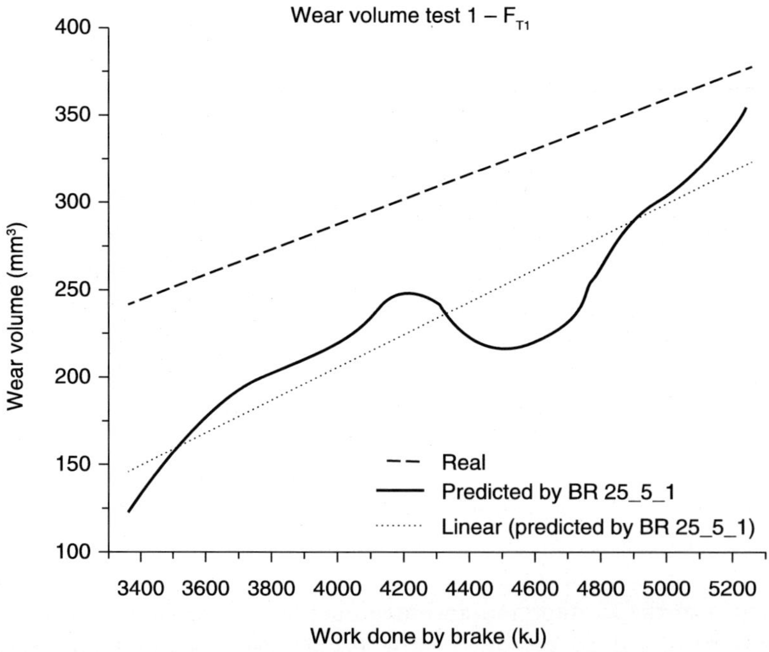

Figure 5.95 Comparison between real and predicted wear volumes of friction material F_{T1} in wear test 1 [208]

content of phenolic resin (17 vol%). Furthermore, this friction material was manufactured at a high moulding pressure (70 MPa) in synergy with a low moulding temperature (155 °C) and low heat treatment time (5 h). The integrity of this brake friction material was significantly affected at elevated temperatures and, consequently, the wear of this friction material was substantially increased (Figure 5.95). This observation could also be connected with the high volume percentage of baryte, used as filler, and the low percentage of reinforcing fibres (see Table 5.41). On the other hand, the friction material F_{T1} showed the best wear performance (see Figure 5.95), as can also be seen in Tables 5.41 and 5.42. The volume percentage of phenolic resin in this friction material was 25. This material was produced using a lower moulding pressure (40 MPa), a higher moulding temperature (170 °C), a higher heat treatment time (12 h), a lower volume percentage of magnesium oxide (used as an abrasive ingredient) (2 vol% versus 8 vol% in the case of F_{T2}) and a higher percentage of reinforcing fibres (21 vol%).

The neural model showed enough flexibility, i.e. generalization capabilities, to predict the variation of the specific wear rate with the formulation and manufacturing conditions of the friction materials in synergy with the work done by the brake and the brake interface temperature. The same neural model (BR 25_5_1) was employed to predict not only the specific wear rates at the end of the wear tests but also changes in the wear volume versus the work done by the brake in wear tests 1, 2 and 3. Figure 5.95 shows a comparison between the real and predicted changes in the wear volume (for the last 15 brake applications) in wear test 1. According Figure 5.95, the predicted values of the wear volume of the friction material F_{T1} converge to the measured value at the end of wear test 1. In order to show the trend of the prediction of wear volume for the friction material F_{T1}, the predicted values of the wear volume were linearized (see Figure 5.95, line 'Linear (predicted by BR 25_5_1)'). Regarding the predicted values of the wear volume of friction material F_{T1} in wear test 1 and the trend of the prediction, it can be concluded that the neural model generalizes well the synergy of the effects of the formulation, manufacturing and operation conditions on the wear of material F_{T1}. The influence of the same formulation and manufacturing parameters of friction material F_{T1} on its wear performance at a higher brake interface temperature of 175 °C (wear test 2) is also predicted (Figure 5.96).

From Figures 5.95 and 5.97, it is evident that the wear volume of friction material F_{T1} increases with an increase in the brake interface temperature. According to Table 5.43, the specific wear rate increases from 0.072 to 0.125 mm^3/kJ as the brake interface temperature increases.

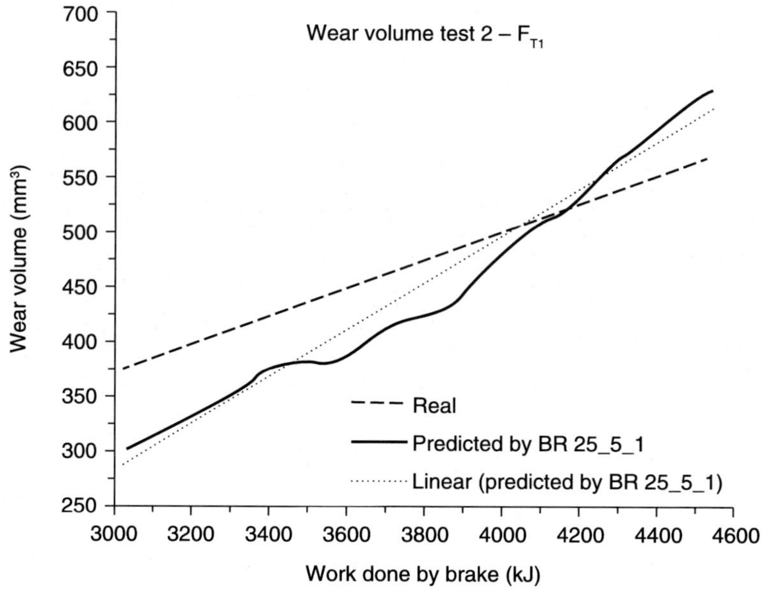

Figure 5.96 Comparison between real and predicted wear volumes of friction material F_{T1} in wear test 2 [208]

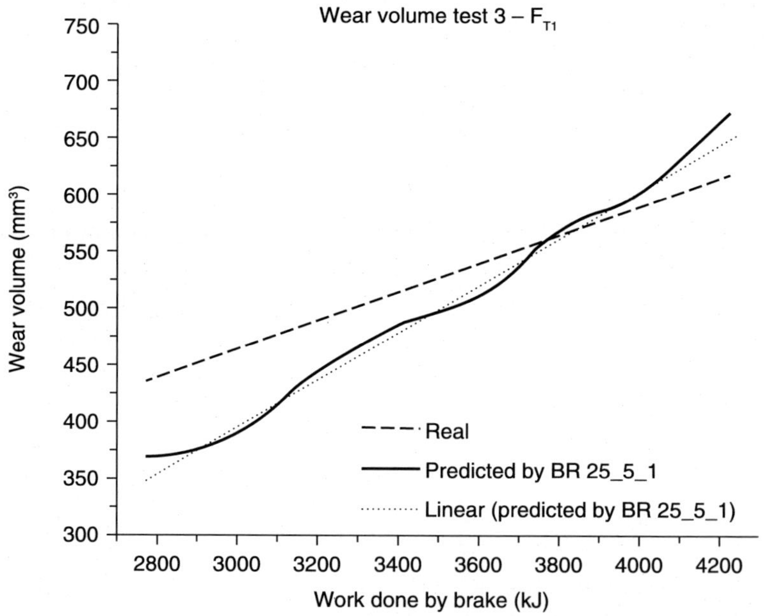

Figure 5.97 Comparison between real and predicted wear volumes of friction material F_{T1} in wear test 3 [208]

The neural model generalizes this influence well, although the trend of the predicted results (see the line 'Linear (predicted by BR 25_5_1)' in Figure 5.97) shows a higher rate of increase in the wear volume versus the work done by the brake. When the brake interface temperature is increased further to 250 °C, the wear volume of friction material F_{T1} is not significantly increased. This material shows relatively stable wear performance as the brake interface temperature increases. The neural model predicts such behaviour well.

From Figure 5.94 and Table 5.43, it can be seen that the friction material F_{T2} shows wear performance different from that of friction material F_{T1}. It is evident that changes in the formulation of the friction material F_{T2} together with changes in the manufacturing conditions cause a drastic increase in the specific wear rate of this friction material. The increase in the wear volume of friction material F_{T2} is particularly evident in wear test 3 (Figure 5.94, for a temperature of 250 °C). Such wear characteristics of the friction material F_{T2} are predicted by the neural model (BR 25_5_1). The predictions of the wear volume of material F_{T2} versus the work done by the brake in wear tests 1, 2 and 3 are shown in Figure 5.98. The same neural model was employed for wear prediction of

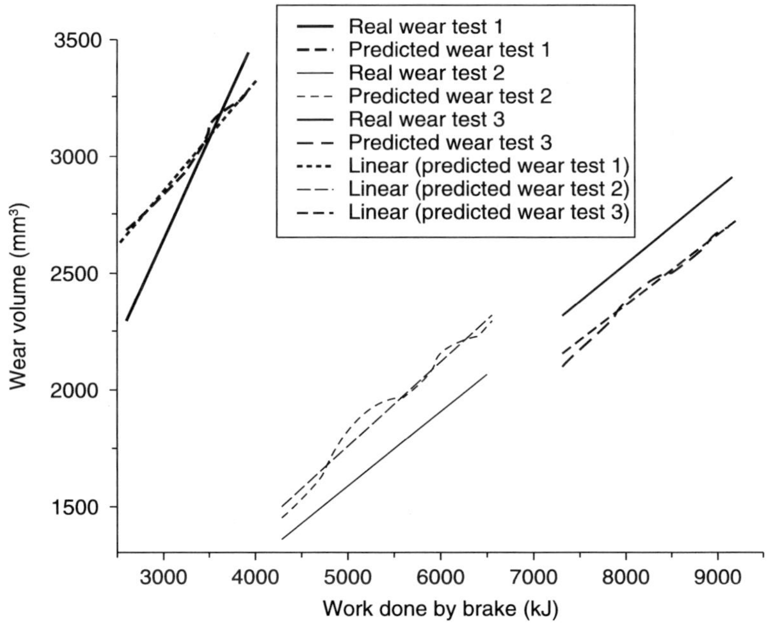

Figure 5.98 Comparison between real and predicted wear volumes of friction material F_{T2} in wear tests 1, 2 and 3 [208]

this type of friction material with a new set of input parameters. According to Table 5.43, the average value of the work done in a brake application is almost the same in all wear tests but the wear volume in wear tests 1, 2 and 3 is influenced in different ways by changes in the brake interface temperature. The abilities of the neural model of friction material wear to predict the wear behaviour of friction material F_{T2} are illustrated in Figure 5.98.

The wear performance of friction material F_{T2} is sensitive to changes in the brake interface temperature. From Figures 5.94 and 5.98, it is evident that the wear of friction material F_{T2} is high but also relatively stable in the range of brake interface temperatures between 100 °C and 175 °C. The wear volume increases when the brake interface temperature is increased to 250 °C. As shown in Figure 5.98, the very complex wear behaviour of friction material F_{T2} in the wear tests demands high generalization capabilities of the neural model. The neural model generalizes well the effect of the synergy of the influencing input factors on the changes in the wear volume in wear tests 1 and 2. From Figure 5.98, it can be seen that for wear test 3, the specific wear rate is predicted to an acceptable level but the gradient of the increase in wear volume versus the work done in the brake applications is not well generalized. In order to test the neural model of the wear of friction materials for a new set of input parameters, the input parameters were selected according to the properties of the friction material F_{T3}. The prediction results are shown in Figure 5.99. Figure 5.99 shows that the wear performance of material F_{T3} is increased when the brake interface temperature increases. The neural model predicts the wear performance of friction material F_{T3} better than it does in the case of friction material F_{T2} (see Figure 5.98). The neural model predicts well the gradient (i.e. slope) of the changes in wear volume versus the work done in the brake applications in all wear tests (Figure 5.99).

In order to illustrate the capabilities of the neural model to generalize the synergistic influence of the input parameters and establish an input/output functional correlation, the prediction capabilities of the neural model are summarized in Figures 5.100 and 5.101. Figure 5.100 shows the predicted specific wear rates, correlated with changes in the manufacturing and brake operation conditions. The predictions of the specific wear rates of friction materials F_{T1}, F_{T2} and F_{T3} are shown over ranges of moulding pressure, moulding temperature and moulding heat treatment time. The neural model of brake friction material wear approximated the synergistic influence of the manufacturing conditions in the three different wear tests, as shown in Figure 5.100. The neural

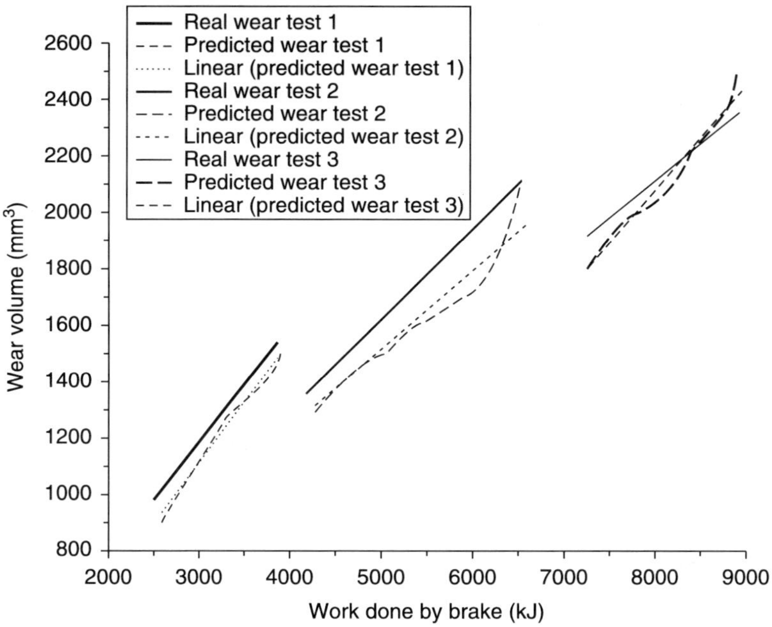

Figure 5.99 **Comparison between real and predicted wear volumes of friction material F_{T3} in wear tests 1, 2 and 3 [208]**

model predicts that increasing the moulding pressure from 40 to 70 MPa, in synergy with a decrease of (i) the moulding time from 11 to 6 min, (ii) the moulding temperature from 170 °C to 155 °C and (iii) the heat treatment time from 12 to 5 h, increases the specific wear rate of the friction material, especially at high brake interface temperatures. The neural-model predictions show that with a lower moulding pressure (40 MPa) and a higher moulding temperature (170 °C) and moulding time, the minimum specific wear rates of these types of friction material can be obtained (see Figure 5.100 again).

The wear performance of the friction materials illustrated in Figure 5.100 is affected not only by changes in the manufacturing and operation conditions; the formulation of the friction materials also needs to be taken into consideration, in synergy with the manufacturing and operation conditions (see Figure 5.101). From Figure 5.101, it can be seen how the specific wear rates are influenced by changing the volume percentages of some of the ingredients in the formulation of the friction materials F_{T1}, F_{T2} and F_{T3}. The synergistic influence of the volume percentages of ingredients such as phenolic resin, baryte, different types

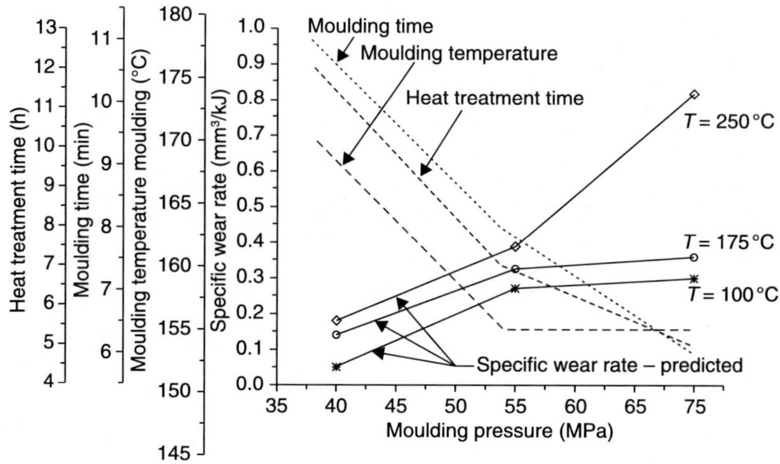

Figure 5.100 Predicted specific wear rate versus influence of the manufacturing and operation conditions of the friction material [208]

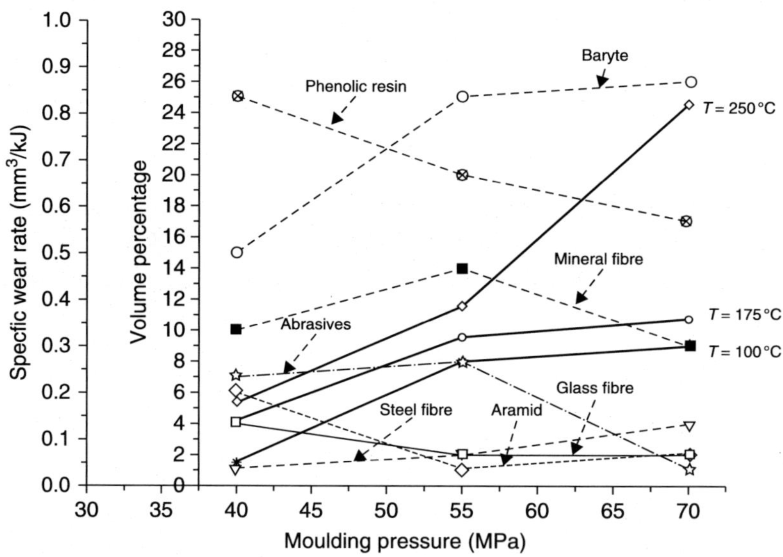

Figure 5.101 Predicted specific wear rate versus influence of the formulation, manufacturing and operation conditions of the friction material [208]

of fibres and abrasives is represented in Figure 5.101. All of these influences were analysed in synergy with the manufacturing and operation conditions. The input/output data mining performed by the neural model shows several different trends in the way in which the formulation, manufacturing and operation conditions of the friction materials affected their wear performance. Figure 5.101 shows that increases in the moulding pressure and the brake interface temperature, in synergy with decreases in the volume percentages of phenolic resin, aramid fibre and glass fibre, cause an increase in the specific wear rate, especially at high brake interface temperatures. This is particularly evident when (i) the moulding pressure exceeds 55 MPa, (ii) the volume percentage of phenolic resin drops below 20%, (iii) the volume percentage of baryte exceeds 25% and (iv) the brake interface temperature reaches 250 °C (see Figure 5.101). For moulding pressures between 40 and 55 MPa, brake interface temperatures between 100 °C and 250 °C, and volume percentages of phenolic resin between 25 and 20% and of baryte between 15 and 25%, all of the friction materials show an increase in the specific wear rate. This is additionally affected by a decrease in the volume percentage of aramid fibre and an increase in the percentage of mineral fibre from 10 to 14 vol% (see Figure 5.101 again). Figure 5.101 shows that the neural model generalizes well the decrease in the specific wear rates of the friction materials over the whole range of brake interface temperatures, under the following conditions: (i) a decrease in the moulding pressure to 40 MPa, (ii) an increase in the moulding time to 11 min, (iii) an increase in the moulding temperature to 170 °C, (iv) an increase in the heat treatment time to 12 h, (v) an increase in the volume percentage of phenolic resin to 25%, (vi) a decrease in the volume percentage of baryte to 15%, (vii) a decrease in the volume percentage of mineral fibre to 10% (the relation between different kinds of fibre in the formulation of the friction material needs to be further optimized), (viii) an increase in the volume percentage of aramid fibre, (ix) a decrease in the volume percentage of steel fibre and (x) an increase in the volume percentage of glass fibre.

5.3.4 Optimization of brake friction material manufacture

Introduction

The performance of the braking system of a motor vehicle is significantly affected by the tribological characteristics of the brake friction pair.

Friction materials are developed to meet comprehensive tribological and mechanical requirements. The most important characteristics of a friction material are related to the level and stability of the friction coefficient during braking and the wear generated by the braking process. The contact situation in a brake pad/disc friction pair is complex and not yet completely understood. The actual contact area between the pad and the disc (in a disc brake) is far from constant; it is very small compared with the total contact area; and it is highly dependent on pressure, temperature, deformation and wear [211, 212]. This situation generates several different braking phenomena. The contact situation is determined mainly by the formulation of the friction material and the manufacturing process employed for its realization. Thus, the development of innovative friction materials is strongly affected by the choice of ingredients, the formulation and the manufacturing parameters.

The manufacturing process of brake friction materials is complex owing to a number of influencing factors. That is why exhaustive technical information about the manufacturing processes of brake linings is seldom found in the literature, although such information is crucial for understanding the physical and tribological properties of these components [213]. Furthermore, the problem of optimization of the manufacturing process of brake friction materials has not been adequately investigated so far. As pointed in [213], brake friction materials are manufactured by hot moulding of a mix under high pressure followed by heat treatments. The hot moulding can be done under high pressure using a positive moulding technology or using 'flash moulding'. In the case of positive moulding, the mix is compressed in a previously heated tool under pressure-controlled conditions.

The most important part of the manufacturing of a brake friction material is related to the compression process. The compression process is interrupted by breathing cycles to expel gases produced in the cavity [214]. It is very important to ensure a constant density of the friction material throughout the whole volume of the final product, in order to ensure the desired friction and wear properties, as well as bonding to the backing plate. The basic problem is related to optimization, i.e. the adaptation of the manufacturing parameters to the formulation of the friction material and to the required friction and wear properties. This is particularly important regarding the relation between the moulding pressure, time and temperature and the content of binder resin in the formulation. This can cause a loss of brake effectiveness at elevated temperatures due to creation of a transfer layer between the contact surfaces. This important characteristic of friction materials is referred to as fade and is of decisive importance in the performance of friction materials.

Obviously, the properties of friction materials are significantly affected by the manufacturing process. The moulding is carried out for less than 15 min under high pressure, and the heat treatment is performed for several hours with no external pressure. The heat treatment is performed to enhance the curing uniformity and to reduce thermal expansion during service. The time, pressure and temperature for each step greatly affect the physical and tribological properties of friction material. It is known that the thermal properties of the binder resin, such as the flow distance, the B-transformation time and the gel time, are closely related to the manufacturing conditions of brake linings [213]. However, the moulding and heat treatment conditions often have to be modified to achieve suitable tribological and physical properties, by changing the types and relative amounts of ingredients in the brake lining. The main problem is how to adjust these manufacturing parameters simultaneously to provide maximum friction and minimum wear in the contact region of the friction pair. Until now, research has focused mainly on the problem of optimizing the formulation of the brake friction material, without significant effort related to the optimization of the manufacturing process applied to the selected formulation [215–218]. The analysis of the synergistic effects of the manufacturing parameters on friction and/or wear is an extremely important and open issue that deserves much attention. The friction and wear performance are two kinds of response of the tribo-system of an automotive brake that are strongly affected by the manufacturing process of the brake friction material. It is very difficult to find an optimal trade-off between the influencing factors. That is why our focus was on correlating the effects of the manufacturing parameters with the wear behaviour of the friction material, in order to optimize the manufacturing process.

An evaluation of the optimal choice of manufacturing process with respect to the wear of friction materials is almost impossible using a traditional analytical approach. The basic deficiency of all wear models describing the response of a brake friction material is related to the objective difficulty of formulating a functional relationship between the formulation, the manufacturing parameters and the operating conditions. Among several experimental-design techniques, the Taguchi method has been applied in a systematic approach to optimizing the design and individuating the manufacturing parameters [213]. The Taguchi analytical methodology was used to investigate how the manufacturing parameters affect the physical properties of brake linings. It was shown that the moulding pressure and the moulding temperature were dominant manufacturing parameters in relation to surface hardness and porosity, respectively [213]. Therefore, the optimization of the manufacturing parameters needs a model that represents their effects on the wear of the brake friction material, as well as a corresponding technique for searching for their optimum values with respect to the goal function and the constraints.

Artificial neural networks were chosen here as an appropriate modelling technique [219–222] for the development of a model representing the effects of the manufacturing process on the wear of friction materials. This model was integrated with genetic algorithms to provide a model with abilities to find the optimal values of the manufacturing parameters. The neural-network model was used for modelling and generalization of the effects of the following factors: (i) the composition of the friction material (volume contents of phenolic resin, fibres, fillers, lubricants and abrasives), (ii) the five most important manufacturing parameters (moulding pressure, moulding time, moulding temperature, heat treatment temperature and heat treatment time) and (iii) the operating conditions of the disc brake (represented by the work done in brake applications) for specific brake interface temperatures (100 and 175 °C). The neural-network model developed was used further for optimization purposes using genetic algorithms. The optimization criterion was based on minimization of the wear volume of the friction material, taking the manufacturing parameters (defined in specified ranges) as variables.

Experimental procedure

The input parameters were defined by the friction material formulation, i.e. the volume percentages of phenolic resin, fibres, fillers, lubricants and abrasives used to obtain the friction material; the manufacturing parameters; and the operating conditions of the disc brake. The operating conditions were expressed in terms of the work done in each brake application and the temperature of the disc brake interface. The wear volume (in mm^3) of the friction material was taken as an output parameter. The wear volume was given by the value measured after a certain number of applications of the disc brake, i.e. after the cumulative work done in all of the brake applications specified in the wear test. The wear volume was calculated based on the linear wear of the friction material, measured at the end of the wear test, by multiplying the average linear wear (in mm) measured on the friction surface by the value of the nominal area of the friction surface. Since the work done in each brake application was known, the cumulative work done in a wear test could be determined (in kJ). The specific wear rate was inferred from the relation between the total wear volume and the cumulative work performed in the wear test (expressed in mm^3/kJ). The disc brake, together with the brake friction material, was tested under strictly controlled conditions with respect to pressure, speed and temperature, using a full-scale inertia dynamometer as a data generator (Figure 5.102). Here, an electric motor (1) drove a carrier (2) and a set of six flywheels (3), located in a protective cage (4), which provided different amounts of inertia (see Figure 5.102 again). The end of the common shaft was firmly joined to the disc (5) of the brake being tested. The stationary part of the brake (the caliper) (6) was connected to a system for measurement of the braking torque (7). The position of this system could be adjusted with an axial slider (8). As can be seen, all of these components were mounted on a common foundation (9).

The friction materials were all tested according to the same testing methodology. The testing methodology was designed in such a way that the applied brake pressure and the initial speed were constant for each braking cycle (initial speed, 90 km/h; applied pressure, 40 bar). The values and procedures indicated here were chosen in order to provide values of the work done in a brake application that corresponded to the most common braking regimes of a passenger car. Since the wear of a friction material is affected by the brake interface temperature, the friction materials were tested at two different brake interface temperatures (100 and 175 °C). These brake interface temperatures were chosen in order to optimize the manufacturing parameters of the brake friction

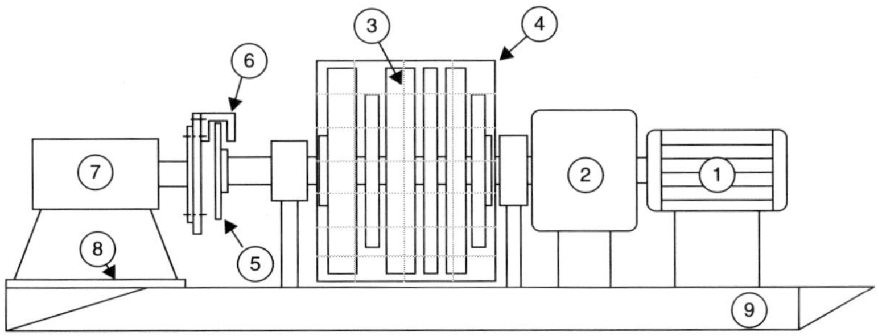

Figure 5.102 **Single-ended full-scale inertia dynamometer [222]**

materials for temperature conditions that correspond to the most common braking regimes of a passenger car. Furthermore, the effects of the brake interface temperature were investigated in order to assess possible differences in the optimal manufacturing parameters for the two brake interface temperatures. The brake interface temperature was measured with a thermocouple inserted into a hole drilled in the inner disc pad at the effective radius of the brake disc (101 mm), located 0.5 mm from the friction surface of the disc pad. Owing to the fact that the temperature sensor was placed inside the friction material, at 0.5 mm from the mating surfaces, there was a difference between the real temperature in the contact region of the disc brake and the measured value; the measured temperature could reasonably be expected to be slightly lower than the real brake interface temperature. The difference will depend on the thermal conductivity of the friction material, which in turn depends on the formulation and, especially, on the quantity of metallic ingredients in the friction material mixture. It was assumed that there was not a significant influence on the measured wear of the friction material as a result of the difference between these two temperatures. Before the wear tests, a gradual transition from transient to steady-state wear conditions of the friction material was provided by burnishing the friction material (with 150 brake applications) using an applied brake pressure of 40 bar and an initial speed of 90 km/h. The influence of the compressibility of the friction material, the thickness of the pad backplate, the coefficient of friction, caliper flexure and the stiffness of the brake disc on the interface pressure distribution were considered as significant [223]. According to [223], in order to ensure consistent performance of a disc brake, the interface pressure distribution must be carefully controlled by designing in a mechanical rigidity compliant with the friction material formulation. Since the wear of the friction materials was optimized for the case of a cold brake (100 °C) and for a relatively low interface temperature in the disc brake (175 °C), the pressure distribution in the friction interface was assumed to be uniform.

The formulation parameters of the friction materials are shown in Table 5.44. The most important issue was related to the percentages of key ingredients such as resin, fibres, fillers, lubricants and abrasives. The formulations of the friction materials were based on types of fibres typically used, namely aramid, mineral, steel and glass fibres. Furthermore, several different friction modifiers were used, such as graphite, friction dust and molybdenum disulphide. Some typical abrasives adopted for this kind of application, namely aluminium oxide, silica and magnesium oxide, were considered in this investigation. The total volume contents of these ingredients were calculated and grouped, as shown in Table 5.44. As can be seen, the influence of the key ingredients was investigated over wide ranges. As shown in Table 5.45, seven different types of friction material were manufactured in order to train the model. Testing data from these friction materials were used to train selected artificial neural networks in order to develop the neural model. The prediction abilities of the model were tested with data stored in the test data set.

As shown in Table 5.45, the friction materials were manufactured under different conditions related to five influencing factors, taken as optimization variables. Owing to the unknown interrelated influences of

Table 5.44 Type and total volume content of ingredients in the friction materials [222]

Ingredient	Training set	Test data set		
	F1–F7	F_{T1}	F_{T2}	F_{T3}
Phenolic resin	15–25	25	17	20
Fibres (aramid, mineral, steel, glass)	15–21	21	17	19
Fillers (iron oxide, baryte, calcium carbonate, brass chips, vermiculite)	20–36	26	36	35
Lubricants (brass powder, copper powder, graphite, friction dust, molybdenum disulphide)	14–21	21	19	17
Abrasives (silica, aluminium oxide, magnesium oxide)	4–11	7	11	9

Table 5.45 Ranges of manufacturing parameters used in shaping the friction materials [222]

Manufacturing parameter	Training set	Test data set		
	F1–F7	F_{T1}	F_{T2}	F_{T3}
Moulding pressure (MPa)	45–70	40	70	55
Moulding temperature (°C)	155–175	170	155	165
Moulding time (min)	5–12	11	6	8
Heat treatment temperature (°C)	200–250	200	200	220
Heat treatment time (h)	5–12	12	5	7

the ingredients in each formulation of the friction materials and the manufacturing parameters on the induced wear, a random distribution was chosen over the selected ranges of formulation and manufacturing parameters (see Tables 5.44 and 5.45).

The friction materials were manufactured as a part of a disc pad assembly by a conventional procedure including dry mixing, preforming, hot moulding and heat treatment (post-curing). Dry mixes were prepared using an electric mixer at 3000 rpm for 8 min and preformed under a moulding pressure of 23 MPa. The preformed moulds of friction materials were press-cured, together with the backing plate, by using a conventional industrial plant for flash moulding, using different pressures, temperatures and times. The disc pad assemblies were post-cured in a conventional electrical curing oven at specified temperatures for specified times (Table 5.45).

Wear modelling

As mentioned in the previous section, input/output data were obtained from the formulation, manufacturing and testing of the selected types of friction material. A neural model of the effects of the formulation, manufacturing and operation parameters of the friction materials was developed taking into consideration 12 different input quantities (Figure 5.103). Modelling the influence of these inputs needed an appropriate neural-network architecture. The architecture of an artificial neural network is described by the number of layers the network has, the number of neurons in each layer, the transfer function of each layer and how the layers are connected to each other.

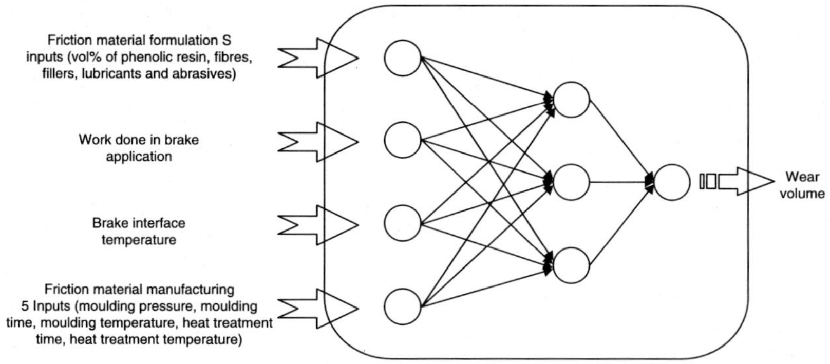

Figure 5.103 **Schematic illustration of the artificial neural network model for wear [222]**

Three different training algorithms (Bayesian regulation, Levenberg–Marquardt and resilient backpropagation) were used for training 10 different architectures of the artificial neural network. A one-layered artificial neural network architecture with eight neurons in the hidden layer and one output neuron, denoted by LM 12 $[8]_1$ 1, trained by the Levenberg–Marquardt algorithm, was found to be the best architecture for input/output functional approximation (Figure 5.102). This model was then integrated with a genetic algorithm to provide the optimization model.

Optimization of manufacturing parameters

Figure 5.104 shows a schematic illustration of how the neural model was used for optimization of the manufacturing parameters. The functional relationship previously established by the neural model between the 12 inputs and one output was incorporated into the new optimization model (Figure 5.103), because the neural model carried knowledge about the complex synergistic influence of the manufacturing parameters of the friction material on its wear. The knowledge stored in the artificial neural model of the wear of friction materials could be extracted through the optimization process. As it was mentioned in the chapter related to genetic algorithms and shown in Figures 5.104 and 5.105, for the specific types of friction material (F_{T1}, F_{T2} and F_{T3}) used for testing, the manufacturing parameters were optimized by using a genetic algorithm, considering the ranges shown in Figure 5.104.

In accordance with Figure 5.105, the genetic algorithm started with an initial population of randomly generated individuals (chromosomes) belonging to the design space and searched for the best input values for

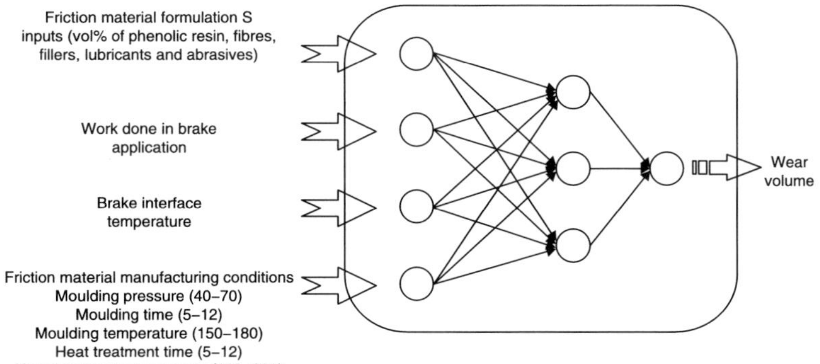

Figure 5.104 Schematic illustration of the neural model used for optimization of manufacturing parameters [222]

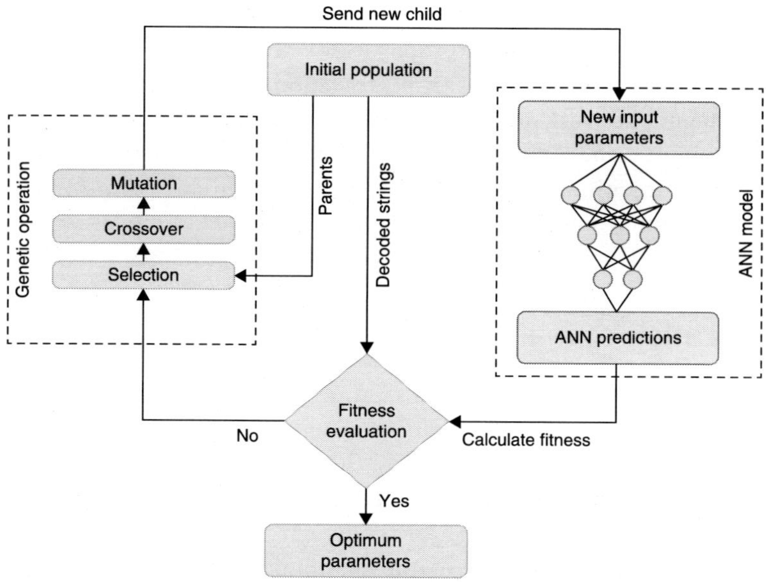

Figure 5.105 Illustration of the hybrid neuro-genetic optimization model used for optimization of manufacturing parameters [222]

the wear model by means of reproduction, crossover and mutation [224–230]. In every evolutionary step, known as a generation, the individuals in the current population were decoded (evaluated) according to a predefined quality criterion, referred to as the fitness or objective function. Figure 5.105 shows a flow chart of the hybrid optimization model developed for the purpose of optimization of the manufacturing parameters. Obviously, the optimization process exploited the ability of the genetic algorithm to optimize the inputs of the neural network related to the manufacturing parameters, using the trained neural model. By simulating the neural model for different inputs related to the formulation, manufacturing and operation parameters, the five manufacturing parameters (Figure 5.104) were optimized according to the fitness function and constraints. The fitness function was defined by the functional relationship between the inputs and the output (Figure 5.104), as established by the neural-network model. This function carried knowledge about the changes in wear versus the effects of the manufacturing parameters, the friction material formulation and the operating conditions. The neural model was able at any time to predict the wear volume of the friction material for the current combination of inputs, including the values of the manufacturing parameters. The fitness function represented by the neural-network model of friction material wear was integrated into the optimization process using a genetic algorithm. The established functional relationship was minimized for inputs taken from the ranges of manufacturing parameters selected (Table 5.44). The optimization constraints were defined according to the ranges in which each manufacturing parameter could be chosen (Figure 5.105) for a specific friction material formulation. The optimization covered the following ranges: moulding pressure between 40 and 70 MPa, moulding time between 5 and 12 min, moulding temperature between 150 and 180 °C, heat treatment temperature between 190 and 220 °C, and heat treatment time between 5 and 12 min.

The process of optimization of the manufacturing parameters was carried out for three different types of friction material, denoted by F_{T1}, F_{T2} and F_{T3}. These friction materials were selected to test the capabilities of the neuro-genetic optimization model to predict and optimize the manufacturing parameters for different brake interface temperatures. Figure 5.106 shows the process of optimization for a brake interface temperature of 100 °C. According to Figure 5.106, a complex synergistic influence of the manufacturing parameters on the wear of friction material F_{T1} clearly exists.

Figure 5.106 shows an almost linear change in the specific wear rate (the wear volume divided by the work done in a brake application, expressed in mm^3/kJ) with increasing moulding pressure. It is evident that

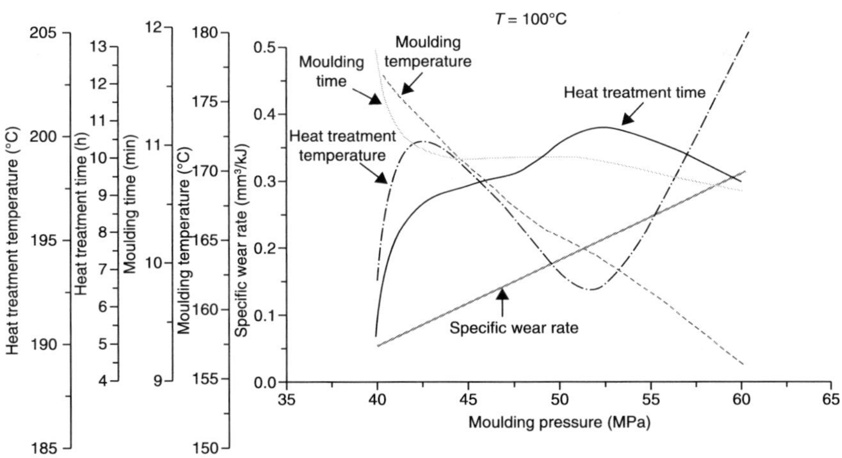

Figure 5.106 **Optimization of the manufacturing parameters of brake friction material F_{T1} for a brake interface temperature of 100 °C [222]**

the moulding pressure and the moulding temperature have a dominant influence on the wear of friction material F_{T1}. According to Figure 5.106, an increase in the moulding pressure from 40 to 62 MPa and a simultaneous decrease in the moulding temperature from 177 to 155 °C, as well as decrease in the moulding time, cause an increase in the specific wear rate from 0.05 to 0.32 mm³/kJ. It is evident that the influence of the heat treatment temperature and the heat treatment time is second order in the case of a brake interface temperature of 100 °C. Thus, the friction material F_{T1}, whose formulation is shown in Table 5.44 (25 vol% of phenolic resin, 21 vol% of fibres, 26 vol% of fillers, 21 vol% of lubricants and 7 vol% of abrasives), should be manufactured according to the conditions shown in Table 5.46. This type of friction material needs a lower moulding pressure, a higher moulding temperature and a higher moulding time to obtain minimum wear under braking conditions corresponding to cold braking (brake interface temperature ≤ 100 °C). Braking regimes with a cold brake are the most common regimes for passenger cars. If different moulding pressures and moulding temperatures are taken into consideration, it can be seen (Figure 5.106) that every increase in the moulding pressure by 10 MPa, with a simultaneous decrease in the moulding temperature by 10 °C, causes an increase in the specific wear rate by three times. This proves that this type of friction material is very sensitive to changes in the manufacturing parameters (the moulding pressure and moulding temperature in this case). Comparing Tables 5.45 and 5.46, it is evident that the neuro-genetic optimization model found optimum values of the moulding pressure, moulding temperature and moulding time that are very close to the parameters used for manufacturing of the friction material F_{T1}. The process of optimization showed that when the moulding temperature was increased from 170 to 177 °C and the heat treatment time was decreased from 12 to 5.2 h, the wear of the friction material could be improved from 0.072 mm³/kJ (the value obtained after testing on the test bench) to 0.05 mm³/kJ.

Table 5.46 **Optimized set of manufacturing parameters for brake friction material F_{T1} for a brake interface temperature of 100 °C [222]**

Moulding pressure (MPa)	Moulding temperature (°C)	Moulding time (min)	Heat treatment temperature (°C)	Heat treatment time (h)
40	177	11.8	193	5.2

As is known, the brake interface temperature has a strong influence on the amount of wear of a friction material. In order to investigate the influence of the brake interface temperature on the wear of brake friction material F_{T1}, the aforementioned optimization model was used to find the manufacturing parameters that would provide a material exhibiting minimum wear at a higher brake interface temperature (175 °C). Since the manufacturing parameters have to be chosen for all operation conditions of a friction material, it is interesting to see the optimal manufacturing parameters in the case of higher brake interface temperatures. The interface temperature in a disc brake under so-called normal braking conditions can be considered to be in the range between 80 and 180 °C. That is why the manufacturing conditions were optimized for the two different temperatures of 100 and 175 °C. Knowing the optimum manufacturing parameters for a brake interface temperature between these two temperatures, it is possible to set the manufacturing parameters to provide minimum wear in this temperature range.

Figure 5.107 shows the process of optimization for the same inputs as in Figure 5.106, but where the brake interface temperature was increased to 175 °C. In Figure 5.107, different influences of the manufacturing parameters on the wear of friction material F_{T1} can be seen. In general, the wear of material F_{T1} is increased. The specific wear rate again increases with an increase in the moulding pressure, but the gradient of the increase is not as high as in the case shown in Figure 5.106. Regarding the

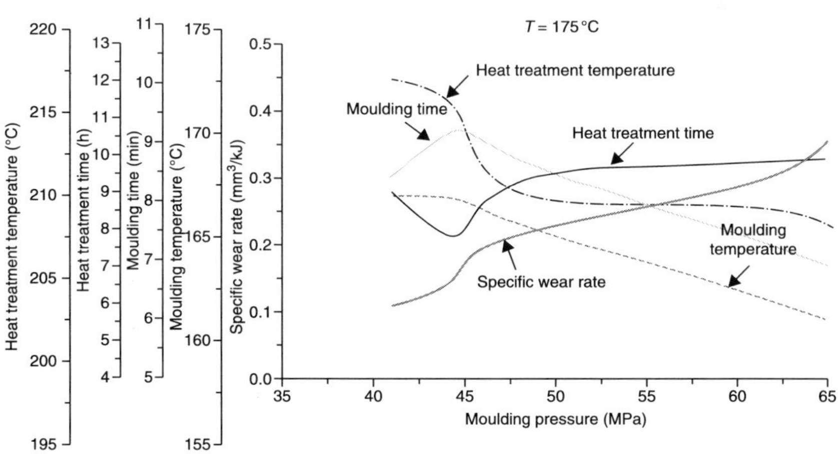

Figure 5.107 **Optimization of the manufacturing parameters of brake friction material F_{T1} for a brake interface temperature of 175 °C [222]**

Table 5.47 **Optimized set of manufacturing conditions for brake friction material F_{T1} for a brake interface temperature of 175 °C [222]**

Moulding pressure (MPa)	Moulding temperature (°C)	Moulding time (min)	Heat treatment temperature (°C)	Heat treatment time (h)
41.5	186	8.4	217	9.2

moulding pressure, its optimum value is almost the same as in the case of a brake interface temperature of 100 °C. Owing to the increase in the brake interface temperature, the minimum specific wear rate is doubled for the same pressure. The specific wear rate, for the same moulding pressure and friction material formulation, increases from 0.05 to 0.125 mm^3/kJ. According to Table 5.47, where the optimum values of the manufacturing parameters are shown, the manufacturing parameters change for the new brake interface temperature, except for the moulding pressure, which is almost the same. Figure 5.107 shows that the moulding pressure, together with the moulding temperature and time, have a dominant influence on the wear of the friction material. The sensitivity of the wear of brake friction material F_{T1} is particularly evident when the moulding pressure changes in the range between 40 and 47 MPa. The heat treatment temperature and time are increased for a brake interface temperature of 175 °C with respect to the case for 100 °C.

The neuro-genetic optimization model was also employed to optimize the manufacturing parameters of the friction material denoted by F_{T2}. The manufacturing parameters of this type of friction material, for which the formulation parameters are shown in Table 5.44, were optimized for both of the above brake interface temperatures. Compared with the friction material F_{T1}, material F_{T2} had lower values of the volume percentages of phenolic resin and fibres (see Table 5.44). The optimization process is shown in Figure 5.108. The optimization model indicates how the specific wear rate is affected by changes in the manufacturing parameters. Figure 5.108 shows that increasing the moulding pressure while decreasing the moulding temperature causes an increase in the specific wear rate from 0.1 mm^3/kJ (for a pressure of 47.6 MPa) to 0.325 mm^3/kJ (for a pressure of 67 MPa). Again, the moulding pressure and temperature, for a brake interface temperature of 100 °C, have a dominant influence on the wear of the friction material F_{T2} (Figure 5.108). Contrary to the manufacturing parameters used for the friction material F_{T2} shown in

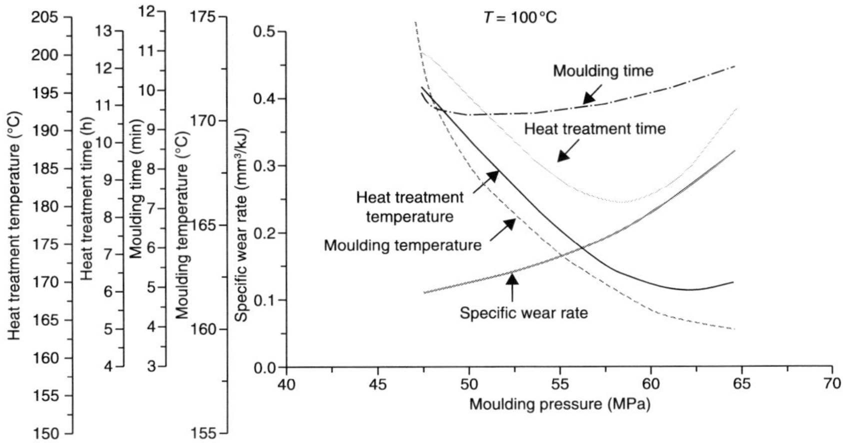

Figure 5.108 **Optimization of the manufacturing parameters of brake friction material F_{T2} for a brake interface temperature of 100 °C [222]**

Table 5.45, the model found a new set of manufacturing parameters (Table 5.48) that could significantly reduce the wear of the friction material. The value of 0.325 mm^3/kJ obtained in testing on the test bench could be reduced further to 0.1 mm^3/kJ by decreasing the moulding pressure from 70 to 47.6 MPa, increasing the moulding temperature from 155 to 175 °C and increasing the moulding time from 5 to 9.8 min.

As was done in the case of friction material F_{T1}, the manufacturing parameters were also optimized for a brake interface temperature of 175 °C. This optimization is shown in Figure 5.109. Again, the minimum wear was obtained for similar values of the moulding pressure, as reported in Table 5.49. The specific wear rate increases when the brake interface temperature is increased for a moulding pressure of 50 MPa. On the other hand, increasing the moulding pressure to 67 MPa does not cause any significant increase in wear with respect to the value obtained for a brake interface temperature of 100 °C. However, increases in the

Table 5.48 **Optimized set of manufacturing parameters for brake friction material F_{T2} for a brake interface temperature of 100 °C [222]**

Moulding pressure (MPa)	Moulding temperature (°C)	Moulding time (min)	Heat treatment temperature (°C)	Heat treatment time (h)
47.6	175.1	9.8	195.1	10

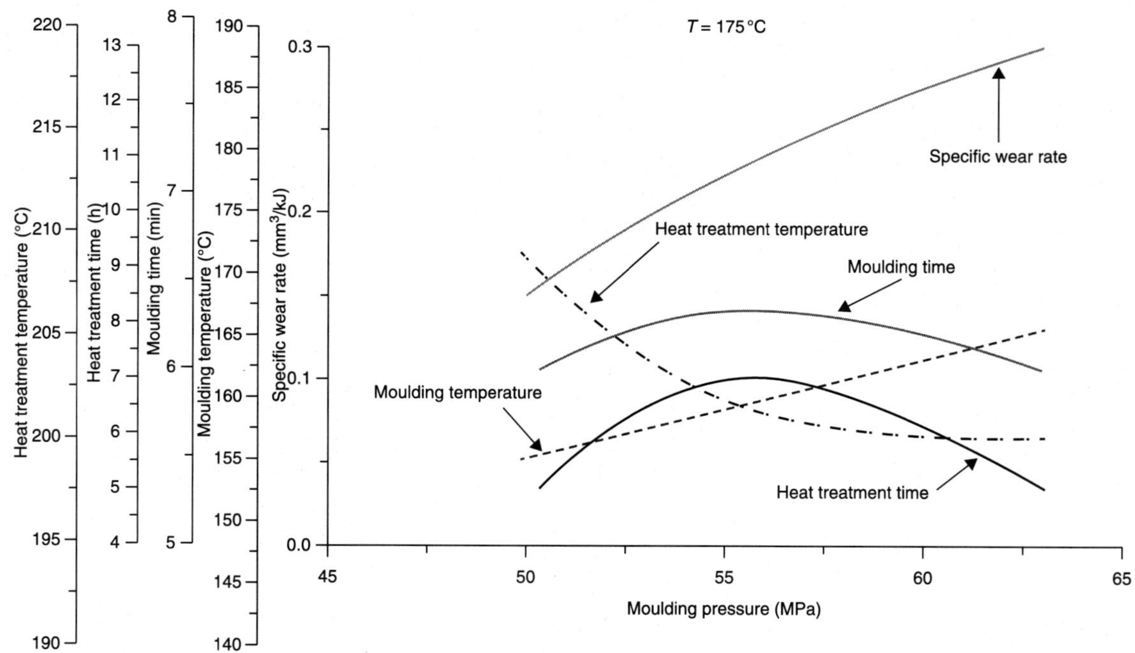

Figure 5.109 Optimization of the manufacturing parameters of brake friction material F_{T2} for a brake interface temperature of 175 °C [222]

Table 5.49 Optimized set of manufacturing parameters for brake friction material F_{T2} for a brake interface temperature of 175 °C [222]

Moulding pressure (MPa)	Moulding temperature (°C)	Moulding time (min)	Heat treatment temperature (°C)	Heat treatment time (h)
50.5	155	6	209	5

moulding pressure and the moulding temperature mostly affect the specific wear rate, causing it to increase. In Figures 5.108 and 5.109, different effects of the moulding temperature on the wear can be seen. In Figure 5.108, a decrease in the moulding temperature causes the wear to increase, but at the higher brake interface temperature (Figure 5.109), a decrease in the moulding temperature causes a decrease in the wear.

Obviously, any changes in the friction material formulation and the brake operation conditions need an adaptation of the manufacturing parameters. In order to prove significant and hard-to-predict effects of the manufacturing parameters on the wear rate of brake friction materials, the friction material denoted by F_{T3} was investigated. This type of friction material was formulated differently from the friction materials F_{T1} and F_{T2}. The neuro-genetic model for wear optimization that we developed was employed to find the optimum set of manufacturing parameters for friction material F_{T3}. As can be seen in Tables 5.45 and 5.46, this type of friction material had a new formulation, and it was manufactured using a different set of manufacturing conditions from those used for the friction materials F_{T1} and F_{T2}. The volume content of fillers was increased to 35% but the content of lubricants was decreased (the volume percentage of phenolic resin was set at 20%). The effects of the change in the manufacturing parameters on the wear are shown in Figure 5.110, for a brake interface temperature of 100 °C. Figure 5.110 shows that an increase in the specific wear rate is again primarily caused by an increase in the moulding pressure, especially for pressures higher than 50 MPa. The rate of increase of the wear is lower than in the case of the friction material F_{T2}. In contrast to the friction material F_{T2}, in the case of material F_{T3}, the effect of the moulding temperature on the specific wear rate is different. An increase in the moulding temperature, in synergy with an increase in the moulding pressure, causes an increase in the specific wear rate (see Figure 5.110). From the same figure, it can be seen that if the moulding pressure is properly chosen, it may be possible to significantly decrease the specific wear rate. The optimized values of the manufacturing parameters are shown in Table 5.50.

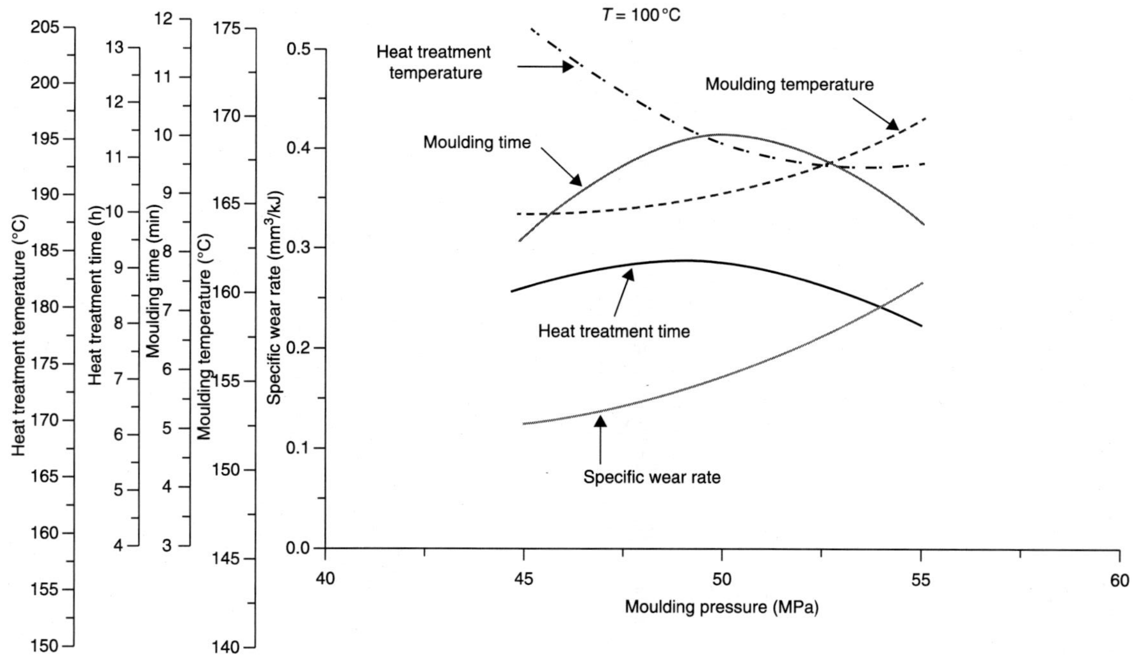

Figure 5.110 Optimization of the manufacturing parameters of brake friction material F_{T3} for a brake interface temperature of 100 °C [222]

Table 5.50 **Optimized set of manufacturing parameters of brake friction material F_{T3} for a brake interface temperature of 100 °C [222]**

Moulding pressure (MPa)	Moulding temperature (°C)	Moulding time (min)	Heat treatment temperature (°C)	Heat treatment time (h)
44	164	8.2	207	8.6

Similarly to the wear behaviour of the friction material F_{T3} at a brake interface temperature of 100 °C, this type of friction material shows an evident sensitivity to an increase in the moulding pressure. A significant increase in the minimum value of the specific wear rate as the brake interface temperature increases from 100 to 175 °C is shown in Figure 5.111. This again highlights the fact that, for this specific formulation of friction material, the dominant influence on the wear is associated with the moulding pressure. The optimized values of the manufacturing parameters are shown in Table 5.51. It can be seen that these values are similar for the two brake interface temperatures.

In order to summarize how the manufacturing parameters influence the wear of the friction materials, the optimal values of the manufacturing parameters are shown in Figure 5.112 for $T = 100$ °C. According to this figure, for these types of friction materials, the moulding pressure should be kept between 40 and 47 MPa, the moulding temperature between 165 and 177 °C, the moulding time between 8.2 and 11.8 min, the heat treatment time between 5.2 and 10 h and the heat treatment temperature between 193 and 207 °C. The moulding pressure and moulding temperature have a dominant influence on the wear of these friction materials. The optimal values of the manufacturing parameters in the case of a brake interface temperature of 175 °C are shown in Figure 5.113.

Table 5.51 **Optimized set of manufacturing parameters of brake friction material F_{T3} for a brake interface temperature of 175 °C [222]**

Moulding pressure (MPa)	Moulding temperature (°C)	Moulding time (min)	Heat treatment temperature (°C)	Heat treatment time (h)
51	168	7.5	210	6.2

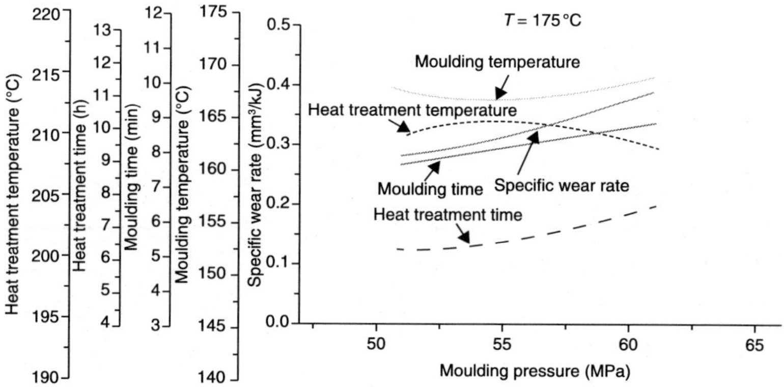

Figure 5.111 Optimization of the manufacturing parameters of brake friction material F_{T3} for a brake interface temperature of 175 °C [222]

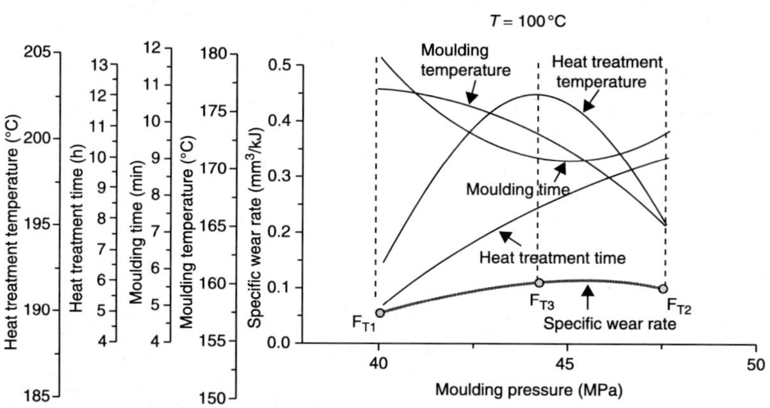

Figure 5.112 Optimal values of manufacturing parameters of brake friction materials F_{T1}, F_{T2} and F_{T3} for a brake interface temperature of 100 °C [222]

Obviously, the minimum wear of the friction materials will be obtained for both brake interface temperatures if the following manufacturing parameters are chosen: a moulding pressure of 40 MPa, a moulding temperature of around 180 °C and a moulding time of 11 min. The wear behaviour of the selected friction materials was not investigated outside

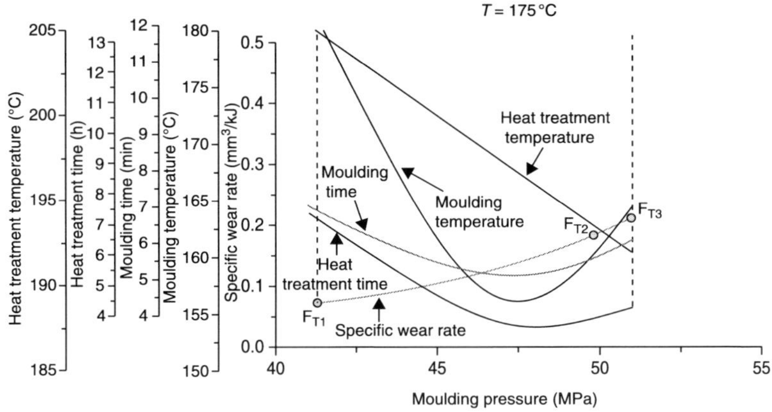

Figure 5.113 **Optimal values of manufacturing parameters of brake friction materials F_{T1}, F_{T2} and F_{T3} for a brake interface temperature of 175 °C [222]**

the boundaries of the moulding-pressure values specified in Figure 5.104. The possibilities for further improvement of the wear of these materials should be investigated further for manufacturing parameters outside these boundaries.

In Figures 5.106–5.113, the moulding pressure is shown as the abscissa so that we can better analyse and illustrate its influence on the wear of the friction materials regardless of changes in the other manufacturing parameters. Figures 5.106 and 5.113 show that, regardless of the manufacturing parameters and the friction material formulation, increasing the moulding pressure causes a significant increase in the wear of the material, especially at elevated brake interface temperatures. Obviously, lower values of the moulding pressure provide better wear behaviour of the friction materials analysed. This can be explained by the complex synergistic influence of the moulding pressure, the volume content of fibres, the volume content of phenolic resin and the characteristics of the phenolic resin (flow distance, B-time, hexa content and free phenol content). This is important because the flow distance represents the melting and flow behaviour of the resin during hot-press moulding and its degassing behaviour. This resin parameter has a dominant influence on the porosity of the friction material and on its final properties, especially at elevated temperatures.

5.3.5 Neuro-genetic optimization of the performance of a brake friction pair

Introduction

Disc brakes are safety-critical automotive components that must satisfy tough cost and environmental requirements [231]. According to [232], drivers often evaluate the performance of automotive braking systems in terms of the brake pedal 'feel'. This is one of the first customer touch points during a driving experience, and as such it can be an important contributor to quality perception and customer appeal [233]. The brake pedal feel depends on the synergistic influence of the performance of the braking system and the driver's subjective perception related to the quality of the brake pedal feel. The brake pedal feel gives the driver a perception of the braking dynamics and braking performance of the vehicle. In this context, it is relevant that the brake pedal feel is significantly affected by the properties of the brake friction material, such as, for instance, the compressibility of the friction material.

Motor vehicle brakes are expected to provide high, stable values of the braking torque under different operating conditions. The operating conditions of a brake are determined by the synergistic influence of the brake actuation pressure, the sliding speed and/or the brake interface temperature [234–238]. It is also well known that the interaction in the brake friction pair determines the performance of a brake. Consequently, brake performance is strongly influenced by the behaviour of the friction material. The sensitivity of the performance of a brake friction material to changes in the brake actuation pressure, in synergy with changes in speed and temperature, is particularly important regarding possible corrections of brake performance during a braking cycle. The control and optimization of brake performance and its sensitivity to changes in speed, pressure and temperature during braking is an issue that needs to be analysed better. In other words, the relevant features of the friction material, as obtained after the development and manufacturing phase, could be enhanced further by identification of the influence of the operating conditions on the final performance of the friction pair. Obviously, this could result in remarkable improvement in the performance exhibited. Nowadays, automotive braking systems are expected to offer stable and at the same time maximum performance under all braking conditions, to exactly match the driver's demands. The operation of motor vehicle brakes needs to be intelligently supported in order to optimize their non-linear behaviour. Owing to the strongly non-linear nature of the braking process, the demands imposed on a braking system and especially on the brakes are very complex and cover a wide range of operating conditions [237, 238]. In the braking systems employed on modern motor vehicles, extreme demands are placed on the friction pair and its tribological performance [239–244].

The most important influence on the sensitivity of brake performance to different braking regimes is the interaction in the friction pair. The brake friction material has a significant influence on the contact situation in the friction pair. As a consequence, the properties of friction materials play a crucial role in the driver's perception of braking performance, especially at elevated brake interface temperatures [242–244]. The properties of the brake friction material influence the performance of disc brakes, i.e. their sensitivity to pressure, speed and temperature during a braking cycle, in several different ways. The brake friction material often critically influences the friction level and its variation during braking, the in-stop friction rise, the hot compressibility [245], and so on. On the other hand, drivers expect a relatively constant level of brake performance in various braking regimes, i.e. the highest level of reliability of the brakes. The level of the braking torque in a braking cycle is a very important issue in the use of a vehicle. It should remain at a stable level in a braking cycle and, at the same time, it should be maximized. However, owing to the frequent, high fluctuations in brake performance that occur, especially the rise in the in-stop braking torque, the stability of brake performance is not ensured. This causes a poor brake pedal feel. Accordingly, the driver obtains confusing feedback about the vehicle dynamics and the performance of the braking system. That is why some level of dynamic control and optimization of the brake performance should be provided during a braking cycle. In other words, a functional relationship between the brake pedal travel and the brake performance needs to be established and optimized, in order to match drivers' demands.

Taking into consideration that very complex and highly non-linear phenomena are involved in the braking process, a sufficiently accurate analytical model of brake performance is almost impossible to obtain [245–249]. The reason for that lies in the fact that the coefficient of friction of a brake friction pair has to be considered as a non-linear parameter. This significantly complicates the process of control of the performance of automotive brakes [250]. That is why intelligent methods need to be introduced into the modelling, control and optimization of highly non-linear processes such as the braking process. The implementation of artificial intelligence techniques should allow the development of intelligent systems that incorporate adaptation and learning. An approach to the optimization of the brake performance of a passenger car, based on a synergy of artificial neural networks and genetic algorithms, is proposed and discussed here. The methodologies and outcomes presented here, in the authors' opinion, could contribute to efforts towards the intelligent optimization of the performance of automotive braking systems and to the creation of 'smart braking' abilities. A hybrid neuro-genetic optimization model was developed for this purpose. This dynamic model was used for control and optimization of the braking torque by adjusting the brake actuation pressure according to the previous and current influences of the sliding speed, the pressure and the temperature. A genetic algorithm was coupled to this model to optimize and adjust the actuation pressure of disc brakes to a level which should provide stable and at the same time maximum performance. The optimization process was done according to the brake pedal travel selected by the driver and the inherent capabilities of the brake for the generation of a braking torque under specific braking conditions. This could be considered as a new, intelligent way to develop an advanced brake assistance system for correcting the current brake performance according to its previous behaviour and the selected brake pedal travel.

Friction pair performance

Braking systems have always been given the highest importance concerning safety issues and, in particular, active safety. Different design and load characteristics of passenger vehicles require different performances of their braking system. The stochastic nature of the operation of friction brakes, i.e. their tribological behaviour, is the main reason why the operation of braking systems needs to be constantly monitored and controlled, depending on the driver's demands and the characteristics of the tyre–road adhesion. That is why the overall capabilities of braking systems are determined mostly by the quality of the operation of the brakes, i.e. the performance of the friction pair in the brakes. Therefore, the performance of automotive brakes results from complex interrelated phenomena occurring in the contact region of the friction pair. These complex braking phenomena are affected mostly by the physicochemical properties of the ingredients of the friction material and the operating regime of the brakes. The performance of brakes is primarily influenced by the contact situation between the cast iron brake disc and the friction material. The contact situation is additionally complicated by the fact that friction materials are complex composites. This situation can be affected in different ways by the wide diversity in the mechanical properties of the ingredients of friction materials [235]. The synergetic effects of all of these ingredients determine the final characteristics of the friction material. Accordingly, brake performance is affected by the interaction in the friction pair and, in turn, the overall performance of the braking system may be dramatically changed. The task of enhancing the performance of automotive brakes, under different operating conditions, is complicated by the fact that the braking process has a stochastic nature. As already mentioned, since very complex, highly non-linear phenomena are involved in the braking process, analytical models of brake operation are difficult or even impossible to obtain. However, in contrast to classical approaches, neural-network modelling may be used to model the complex, non-linear, multi-dimensional functional relationships between the brake inputs and outputs. In order to illustrate the complex changes in the brake factor C, a 3D plot of changes in the performance of a brake is shown in Figure 5.114. There is no doubt that the variations of the brake factor are significantly affected by the friction material used. This is especially evident under high brake actuation pressure. Regarding the synergistic pressure–speed influence, in

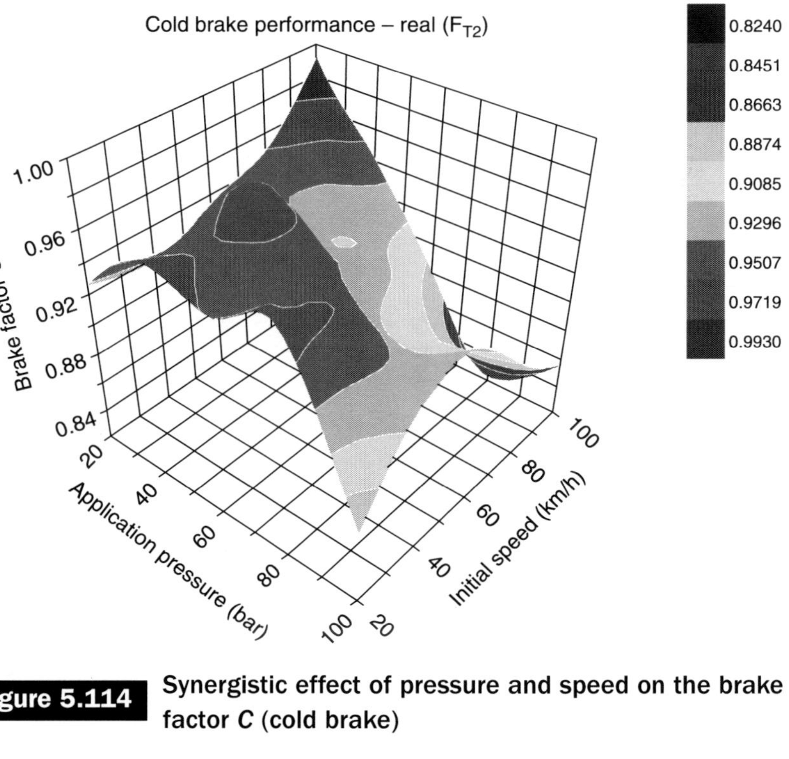

Figure 5.114 **Synergistic effect of pressure and speed on the brake factor *C* (cold brake)**

the case of a cold brake, the brake actuation pressure needs to be precisely controlled during the braking cycle in order to provide maximum performance. That is why it is important to develop an optimization model able to tune the brake actuation pressure to the level which provides the wanted brake performance.

From Figure 5.115, it can be seen that the best performance of the brake is reached between 40 and 60 bar for all values of the initial speed between 20 and 100 km/h. Furthermore, a decrease in the brake performance is evident when the brake application pressure is greater than 60 bar. Therefore, it is important to know, i.e. identify, the sensitivity of the brake performance to different operation regimes (pressures, speeds and temperatures). This could be used for better selection of the brake actuation pressure according to the driver's demands. For instance, during normal braking (initial speed 60 km/h; the brake actuation pressure selected by the driver corresponds to 30 bar), the pressure selected by the driver could possibly be increased from 30 to 40 bar by a brake controller if the deceleration of the vehicle does not correspond to that desired by the driver. The decision about which value of the pressure should be selected by the controller depends on the driver's demand and the actual deceleration of the vehicle. In the case of panic braking, for a vehicle speed of 100 km/h

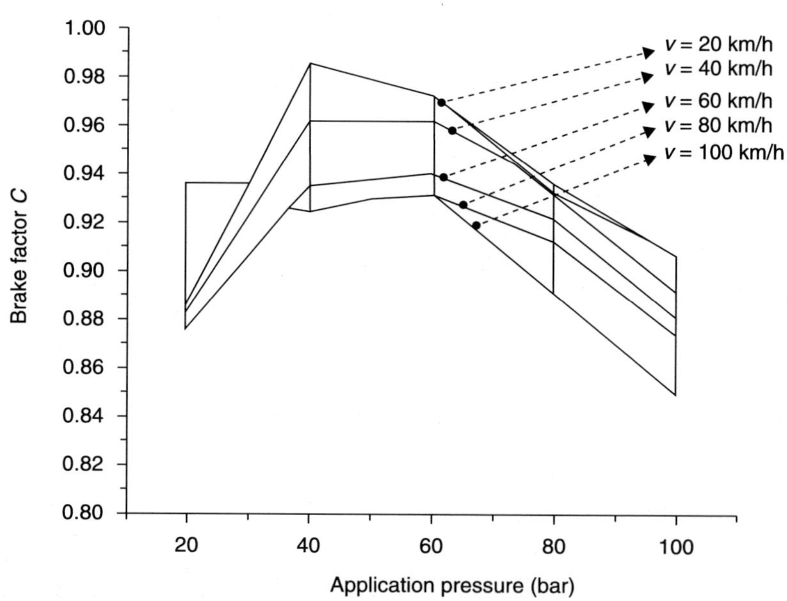

Figure 5.115 **Changes in brake factor *C* versus synergistic pressure–speed influence [236]**

and a pressure selected by the driver of 60 bar, the brake actuation pressure would be increased to the maximum level. Obviously, looking at Figure 5.115, we can see that the brake performance could be maximized if the brake actuation pressure was harmonized with the current performance of the brake friction pair. As illustrated in Figure 5.115, for the same actuation pressure selected by the driver, the brake controller can sometimes increase or even decrease the brake actuation pressure further, depending on the braking strategy and the real brake performance. It is important that the brake actuation pressure is harmonized with the decrease in speed and the increase in brake interface temperature during a braking cycle.

In order to find the correlation between the brake pedal travel, or in other words the selected brake actuation pressure, and the induced brake performance, appropriate data have to be provided for the purpose of modelling. That is why a disc brake from a passenger car was previously tested using a single-ended full-scale inertia dynamometer. This dynamometer was used in order to provide data under strictly controlled test conditions, in accordance with the selected testing methodology. This is very important because the characteristics of a friction material critically affect the overall brake performance and its changes with speed, pressure and temperature during a braking cycle. This effect is primarily caused by the formulation of the friction material and its compressibility, which affects, for example, the in-stop friction behaviour (Figure 5.116).

As mentioned, the important properties of future smart brakes are related to the possible control, i.e. stabilization and maximization, of brake performance versus the brake pedal travel selected by the driver and the current braking conditions. In the study presented here, the influence of several interrelated parameters, such as the vehicle speed, the brake actuation pressure and the brake interface temperature, during a braking cycle was investigated in terms of optimization of the brake actuation pressure. A new model for the optimization of the brake actuation pressure of a passenger car was developed. This optimization model provides different braking strategies according to the wanted brake performance. The model is able to optimize the value of the actuation pressure of disc brakes during a braking cycle versus the brake pedal travel selected by the driver and the tribological behaviour of the brakes. The main criterion for the optimization of the brake actuation pressure is the stabilization and, depending on the requirements of the driver, the maximization of the performance (braking torque) of the friction pair.

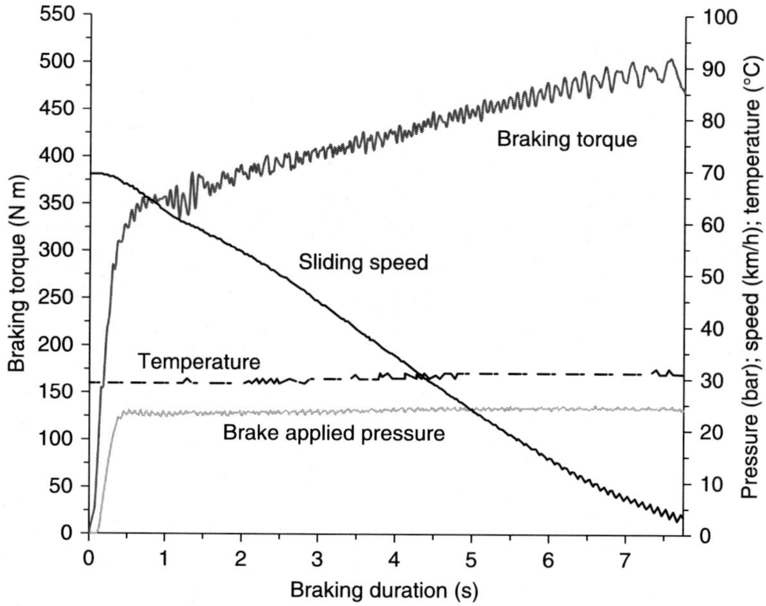

Figure 5.116 **Changes in braking torque during a braking cycle (low brake actuation pressure)**

Neuro-genetic optimization of brake performance

The diversity and substantial variation of the braking torque under the synergistic influence of the pressure, sliding speed and brake interface temperature need to be better controlled in order to provide stable and at the same time maximum braking performance. The non-homogeneous temperature distribution in the contact region of the friction pair, variations in the coefficient of friction, noise and induced vibrations are the main negative issues in the interaction of brake friction surfaces, particularly at high sliding speed. These negative influences could be reduced through optimization of the brake actuation pressure during the braking cycle, as proposed here. The optimization of brake performance by adjusting the brake actuation pressure during a braking cycle could provide several different braking strategies for the braking system. Using the proposed approach, the desired brake performance could be achieved in several different ways.

The maximum performance depends on the current tribological behaviour of the brake friction pair, determined by the interaction between the friction material and the brake disc. In the cases considered here, the goal of the optimization procedure might be related only to the maximization of the brake performance without stabilization of the braking torque during the braking cycle (i.e. brake assistance), if that is all that is needed. A second approach could be to optimize the brake actuation pressure to provide not only the maximum braking performance but also to try to stabilize the braking torque. This could be important when significant fluctuations in the brake performance occur during a braking cycle. Stabilization of the braking torque could be achieved by correcting the brake actuation pressure taking account of the current tribological behaviour of the brake, i.e. correcting the sensitivity of the brake performance to the current braking regime. This is important because the second approach could eliminate the increase in braking torque that occurs with the decrease in speed during a braking cycle, for example. This negative effect of brake operation could be avoided or suppressed by adjusting the brake actuation pressure during the braking cycle, especially at the end of the cycle.

Figure 5.117 shows the process of optimization of the brake actuation pressure versus the real pressure selected by the driver. The optimization was based on the previously developed neural-network model of brake performance in different braking regimes, simulated using the single-ended full-scale inertia dynamometer. According to Figure 5.117, variations in the brake operation conditions cause variations in the braking torque over a wide range, related to the synergistic influence of

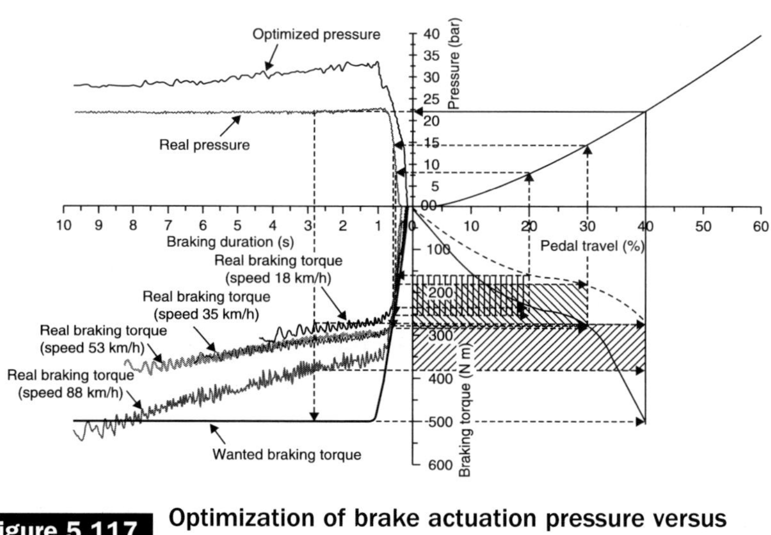

Figure 5.117 **Optimization of brake actuation pressure versus braking torque (pedal travel 40%)**

the pressure, speed and temperature during a braking cycle. Obviously, for the same brake pedal travel of 40% and a real brake actuation pressure of 22 bar (see Figure 5.117 again), the minimum performance is obtained for an initial speed of 18 km/h. The maximum performance is reached when the initial speed is 88 km/h. The maximum braking torque varies between 300 N m (at 18 km/h) and 550 N m (at 88 km/h). Accordingly, the brake performance shows a serious sensitivity to changes in the sliding speed during a braking cycle, in the case of a cold brake. Consequently, if the driver wants the same braking performance for the same pedal travel but different initial speeds, the brake pedal travel must always be changed. Owing to the pressure–speed sensitivity of the friction pair, a number of negative effects can occur during braking. This situation could change drastically when there is an elevated temperature at the brake interface.

The importance of harmonization of the brake performance with respect to the brake pedal travel selected by the driver can be seen by considering two specific cases: (i) in the case of a very high initial braking torque (or friction coefficient), drivers often feel that the brakes are very 'grabby', and (ii) in the case of a low initial braking torque, drivers can feel an absence of braking. For this reason, the assessment of brake pedal feel should take account of the so-called in-stop friction performance, because significant fluctuations in the braking torque can occur. These fluctuations in the braking torque reflect the complex situation in the friction pair contact. Figure 5.117 shows that the in-stop friction performance of the brake considered here is not good. According to [232], the optimal in-stop performance can be defined as a 10–20% increase in friction, i.e. braking torque. However, according to Figure 5.117, a 45% in-stop friction performance is obtained for an initial speed of 88 km/h. A high in-stop friction performance results in a grabby pedal feel at the end of braking and a soft pedal feel at the start of braking. This causes an unpleasant feel of the braking performance of the vehicle, and often the brake is prone to noise generation at the end of braking. Moreover, the braking distance can be drastically increased, which affects the active safety of the vehicle. That is why the initial value of the braking torque as well as its value at the end of braking should be corrected.

It is important to emphasize that dynamic adaptation of the brake performance has to be done during a braking cycle. Based on this, the wanted brake performance can be defined in different ways. This is important particularly in the case when the braking torque has to be adapted to the tyre–road adhesion conditions. As shown in Figure 5.117, the brake actuation pressure was optimized with respect to changes in the value of braking torque denoted as 'wanted braking torque'. Some results of the optimization process are shown in Figure 5.117. It is evident, for the selected brake pedal travel of 40% and real applied brake pressure of 22 bar generated, that there is a significant variation in the brake performance with changes in speed. In the case where the maximum performance for this pedal travel was set at the braking-torque level denoted as 'wanted braking torque', the neuro-genetic optimization model suggested the brake actuation pressure during the braking cycle denoted as 'optimized pressure' in Figure 5.117. As shown in this figure, the neuro-genetic optimization model, based on a dynamic model of brake performance, was used to optimize the brake actuation pressure during the braking cycle. The optimized variation of the brake actuation pressure shown in Figure 5.117 modulates the real brake actuation pressure in such a manner that the initial pressure is increased to 34 bar and is slightly decreased at the end of the braking cycle.

The braking situation described above is related to the most common braking situation, of the so-called light braking condition (with a low value of the brake actuation pressure). As has been shown, the brake performance under such conditions is very sensitive to changes in the sliding speed. In order to analyse a braking situation which was more critical from the point of view of the braking distance and the brake pedal feel, the brake pedal travel was selected as 62% (Figure 5.118). As

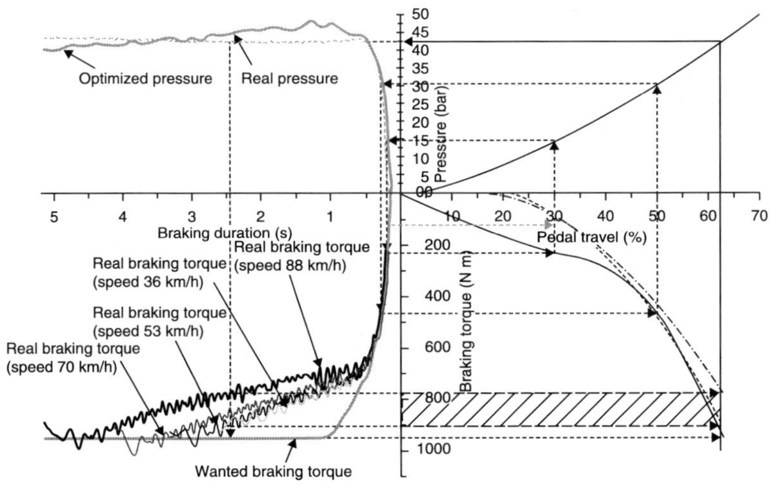

Figure 5.118 Optimization of brake actuation pressure versus braking torque (pedal travel 62%)

shown in Figure 5.118, the brake actuation pressure generated by the driver was 42 bar (real pressure). The maximum range of variation of the real braking torque, for different initial speeds, was around 400 N m. This common braking situation should be avoided because it leads to a poor brake pedal feel and a longer braking distance. The minimum braking torque is obtained for an initial speed of 88 km/h, in contrast to the previous case, and the maximum braking torque is obtained for an initial speed of 36 km/h. As described in the case of a brake pedal travel of 40% (see Figure 5.117), the brake actuation pressure was optimized again. Slight changes in the brake actuation pressure at the start of the braking cycle can partially suppress or even eliminate the differences in braking torque caused by the speed sensitivity of the friction pair. If the applied brake pressure is increased to 47 bar at the start of braking, the responsiveness of the brake is improved and the change in braking torque is stabilized to some extent. It is evident from Figure 5.118 that the brake pedal feel is changed with respect to the previous situation shown in Figure 5.117. It can be seen that for values of the brake pedal travel up to 30%, the brake generates low braking-torque values (see Figure 5.118). For values of the brake pedal travel between 30 and 50%, the responsiveness of the brake is substantially increased. In the range of brake pedal travel between 50 and 70%, the responsiveness of the brake is also increased. In order to correct the brake performance at the end of the braking cycle in the sense of stabilization, the brake pressure was slightly decreased at the end of the cycle. Furthermore, the braking time and braking distance can be shortened by taking the tyre–road adhesion conditions into account. According to Figures 5.117 and 5.118, the brake performance can be controlled by correcting the brake actuation pressure selected by the driver. Furthermore, a unique brake pedal feel can be established with respect to the brake pedal travel. It is important to note that the brake performance can be stabilized better in this case for brake actuation pressures more than 45 bar. This can be explained as an effect of the compressibility of the friction material.

5.4 Conclusions

Composite materials play a crucial role in several engineering areas. The design and development of composite materials is a complex process because composite materials must be formulated and manufactured in such a way that they provide the required in-service performance. Since a number of ingredients can be used in the formulation of a composite material, whose properties can be affected in different ways by the manufacturing process, there is a crucial issue related to the investigation of the possibilities for modelling, prediction and optimization of the performance of composite materials. This is especially important for those materials for which the performance depends strongly on operation conditions that vary over a wide range. The main problem is how to model these influencing factors with respect to the performance of the composite material. Having a model of the behaviour of a composite material under specific operation conditions is important in order to provide preconditions for the prediction and optimization of its performance. Analytical models of the behaviour of composite materials under different operation conditions are difficult or even impossible to obtain. In this chapter, it has been shown how the performance of composite materials can be improved using the abilities of soft computing. These abilities have been explored in order to establish non-linear functional relationships between input and output data. Experimental and numerical data were used in the work described in this chapter for that purpose. It can be concluded that the soft computing techniques presented here offer acceptable possibilities for the process of modelling, prediction and optimization of the performance of composite materials. It has been shown that important aspects of the performance of composite materials can be modelled and optimized using the integration of soft computing techniques with other computational methods. This was illustrated using composite brake friction materials and thermoset matrix composites.

Computational models based on FDM, FEM or FVM methods can be used effectively to predict relevant aspects related to the manufacturing process of a generic thermoset matrix composite material. The reliability of the numerical results has been highlighted with respect to autoclave curing, pultrusion, and the isothermal and non-isothermal–reactive impregnation stages of single-scale and dual-scale porous media in liquid composite moulding processes, by comparison with available experimental data. As a consequence, these models can be considered as excellent candidates for creating the data sets needed for the training, testing and validation of artificial neural networks. Undoubtedly, this procedure could drastically compress the time and cost compared with a full experimental approach, by reducing the experimental effort to the validation of numerical models. What is more, the combination of these models with optimization algorithms, such as genetic algorithms, simplex-type methods or simulated annealing algorithms, allows one to individuate optimal manufacturing conditions with respect to a product- or process-oriented fitness function, as reported in relation to the design of an autoclave thermal cycle and of the heating profile in a conventional pultrusion process.

Brake friction materials usually contain more than 20 different ingredients, and highly complex tribological interactions may take place during the interaction of the friction pair. An approach to modelling the influence of all relevant factors, such as the formulation of the brake friction material, its manufacturing conditions and the brake operating regime, has been investigated. It was shown that computer-based models provided by artificial neural networks may help in resolving the complex task of prediction of the performance of brake friction materials under different operation conditions (cold performance and fading performance, for instance). A neural model of the wear of brake friction materials was also developed for three different wear tests. Based on the models developed for the friction and wear behaviour of brake friction materials, the formulation and manufacturing conditions could be optimized. This was shown for different formulation and manufacturing conditions of brake friction materials.

Based on the results presented in this chapter, it can be said that soft computing techniques are a helpful tool for mining experimental data and searching for patterns in the behaviour of composite materials under prescribed operation conditions. In this regard, the experimental phase is a critical step for the later successful use of soft computing techniques. The procedure for collection of experimental data needs to be strictly controlled in terms of the formulation and manufacture of the composite material, as well as of its testing, using an appropriate data generator. Furthermore, soft computing can be used not only for the purpose of optimization of composite material manufacturing processes but also as a technique for dynamic optimization of the performance of a friction pair, as was shown in Section 5.3.5 in relation to the optimization of the performance of a disc brake friction pair during a braking cycle.

5.5 References

[1] Aleksendrić D., Cirović V. Smart brakes – neuro-genetic control of brake actuation pressure, in Bacciga A., Nalito R. (eds), *Recent Advances in Artificial Intelligence Research*, pp. 85–102, Nova Science Publishers, New York (2013).

[2] Gutowski T.G. *Advanced Composites Manufacturing*, Wiley Interscience, New York (1997).

[3] Pillai K.M. Modeling the unsaturated flow in liquid composite molding processes: a review and some thoughts, *Journal of Composite Materials*, 38(23), 2097–118 (2004).

[4] Advani S.G. *Flow and Rheology in Polymer Composites Manufacturing*, Elsevier Science (1994).

[5] Tan H., Pillai K.M. Multiscale modeling of unsaturated flow in dual-scale fiber preforms of liquid composite molding III: reactive flows, *Composites Part A: Applied Science and Manufacturing*, 43(1), 29–44 (2012).

[6] Spaid M.A.A., Phelan Jr F.R. Modeling void formation dynamics in fibrous porous media with the lattice Boltzmann method, *Composites Part A: Applied Science and Manufacturing*, 29, 749–55 (1998).

[7] Park C.H., Lebel A., Saouab A., Bréard J., Lee W.I. Modeling and simulation of voids and saturation in liquid composite molding, *Composites Part A: Applied Science and Manufacturing*, 42(6), 658–68 (2011).

[8] Loos A.C., Springer G.S. Curing of epoxy matrix composites, *Journal of Composite Materials*, 17(2), 135–69 (1983).

[9] Alavi-Soltani S., Sabzevari S., Koushyar H., Minaie B. Thermal, rheological, and mechanical properties of a polymer composite cured at different isothermal cure temperatures, *Journal of Composite Materials*, 46(5), 575–87 (2012).

[10] Wisnom M.R., Gigliotti M., Ersoy N., Campbell M., Potter K.D. Mechanisms generating residual stresses and distortion during manufacture of polymer–matrix, *Composite Structures*, 37, 522–529 (2006).

[11] Msallem Y.A., Jacquemin F., Boyard N., Poitou A., Delaunay D., Chatel S. Material characterization and residual stresses simulation during the manufacturing process of epoxy matrix composites, *Composites: Part A: Applied Science and Manufacturing*, 41, 108–115 (2010).

[12] Davé R. A unified approach to modeling resin flow during composite processing, *Journal of Composite Materials*, 24(1), 22–41 (1990).

[13] Simacek P, Advani S.G. A numerical model to predict fiber tow saturation during liquid composite molding, *Composites Science and Technology*, 63(12), 1725–36 (2003).

[14] Tan H., Pillai K.M. Multiscale modeling of unsaturated flow in dual-scale fiber preforms of liquid composite molding I: isothermal flows, *Composites Part A: Applied Science and Manufacturing*, 43(1), 1–13 (2012).

[15] Tan H., Pillai K.M. Multiscale modeling of unsaturated flow in dual-scale fiber preforms of liquid composite molding II: non-isothermal flows, *Composites Part A: Applied Science and Manufacturing*, 43(1), 14–28 (2012).

[16] Kaviany M. *Principles of Heat Transfer in Porous Media*, Springer, New York (1995).

[17] Borchardt H.J., Daniels F. The application of differential analysis to the study of reaction kinetics, *Journal of the American Chemical Society*, 79, 41–6 (1956).

[18] Kissinger H.E. Reaction kinetics in differential thermal analysis, *Analytical Chemistry*, 29, 1702–6 (1957).

[19] Ozawa T. Kinetic analysis of derivative curves in thermal analysis, *Journal of Thermal Analysis*, 2, 301–4 (1970).

[20] Horie K., Hiura H., Sawada M., Mita I., Kambe H. Calorimetric investigation of polymerization reaction. III. Curing reaction of epoxides with amines, *Journal of Polymer Science, Part A-1*, 8, 1357–72 (1970).

[21] Chern C.S., Poehlein G.W. A kinetic-model for curing reactions of epoxides with amines, *Polymer Engineering and Science*, 27, 788–795 (1987).

[22] Castro J. M., Macosko C.W. Kinetics and rheology of typical polyurethane reaction injection molding, *SPE Technical Papers*, 19, 434–8 (1980).

[23] Ngo N.D., Mohan R.V., Chung P.W., Tamma K.K., Shires D.R. Recent developments encompassing non-isothermal/isothermal liquid composite molding process modeling/analysis: physically accurate, computationally effective, and affordable simulations and validations, *Journal of Thermoplastic Composite Materials*, 11(6), 493–532 (1998).

[24] Karkanas P.I., Partrige I.K. Cure modeling and monitoring of epoxy/amine resin systems. I. Cure kinetics modeling, *Journal of Applied Polymer Science*, 77, 1419–31 (2000).

[25] Karkanas P.I., Partrige I.K. Cure modeling and monitoring of epoxy/amine resin systems. II. Network formation and chemoviscosity modeling, *Journal of Applied Polymer Science*, 77, 2178–88 (2000).

[26] Pascault J.P., Williams R.J.J. Glass transition temperature versus conversion relationships for thermosetting polymers, *Journal of Polymer Science, Part B: Polymer Physics*, 28, 85–95 (1990).

[27] Di Benedetto A.T. Prediction of the glass transition temperature of polymers: a model based on the principle of corresponding states, *Journal of Polymer Science, Part B: Polymer Physics*, 25, 1949–69 (1987).

[28] Ruiz E., Trochu F. Numerical analysis of cure temperature and internal stresses in thin and thick RTM parts, *Composites Part A: Applied Science and Manufacturing*, 36, 806–26 (2005).

[29] Bogetti T.A., Gillespie Jr J.W. Process-induced stress and deformation in thick section thermoset composite laminates, *Journal of Composite Materials*, 26(5), 626–60 (1992).

[30] White S.R., Hahn H.T. Process modeling of composite materials: residual stress development during cure. Part I. Model formulation, *Journal of Composite Materials*, 26(16), 2402–22 (1992).

[31] Johnston A., Vaziri R., Poursartip A. A plane strain model for process-induced deformation of laminated composite structures, *Journal of Composite Materials*, 35(16), 1435–69 (2001).

[32] Johnston A. An integrated model of the development of process-induced deformation in autoclave processing of composites structures, PhD thesis, The University of British Columbia, Vancouver (1997).

[33] Kwon Y.W., Allen D.H., Talreja R. *Multiscale Modeling and Simulation of Composite Materials and Structures*, Springer Science, New York (2008).

[34] Chung P.W., Tamma K.K., Namburu R.R. Asymptotic expansion homogenization for heterogeneous media: computational issues and applications, *Composites Part A: Applied Science and Manufacturing*, 32(9), 1291–1301 (2001).

[35] Steinhauser M. *Computational Multiscale Modeling of Fluids and Solids: Theory and Applications*, Springer Science, New York (2008).

[36] Ngo N.D., Tamma K.K. Microscale permeability predictions of porous fibrous media, *International Journal of Heat and Mass Transfer*, 44, 3135–45 (2001).
[37] Song Y.S., Youn J.R. Asymptotic expansion homogenization of permeability tensor for plain woven fabrics, *Composites Part A: Applied Science and Manufacturing*, 37, 2080–7 (2006).
[38] Trias D., Costa J., Mayugo J.A., Hurtado J.E. Random models versus periodic models for fibre reinforced composites, *Computational Materials Science*, 38, 316–24 (2006).
[39] Chen X., Papathanasiou T.D. Interface stress distributions in transversely loaded continuous fiber composites: parallel computation in multi-fiber RVEs using the boundary element method, *Computational Materials Science*, 64, 1101–14 (2004).
[40] Bechtold G., Ye L. Influence of fibre distribution on the transverse flow permeability in fibre bundles, *Composites Science and Technology*, 63, 2069–79 (2003).
[41] Chen X., Papathanasiou T.D. On the variability of the Kozeny constant for saturated flow across unidirectional disordered fiber arrays, *Composites Part A: Applied Science and Manufacturing*, 37, 836–46 (2006).
[42] Chen X., Papathanasiou T.D. Micro-scale modeling of axial flow through unidirectional disordered fiber arrays, *Composites Science and Technology*, 67, 1286–93 (2007).
[43] Chen X., Papathanasiou T.D. The transverse permeability of disordered fiber arrays: a statistical correlation in terms of the mean nearest interfiber spacing, *Transport in Porous Media*, 71, 233–51 (2008).
[44] Vaughan T.J., McCarthy C.T. A combined experimental–numerical approach for generating statistically equivalent fibre distributions for high strength laminated composite materials, *Composites Science and Technology*, 70, 291–7 (2010).
[45] Melro A.R., Camanho P.P., Pinho S.T. Generation of random distribution of fibres in long-fibre reinforced composites, *Composites Science and Technology*, 68, 2092–102 (2008).
[46] Morisita M. Measuring of the dispersion and analysis of distribution patterns, *Memoirs of the Faculty of Science, Kyushu University, Series E: Biology*, 2, 215–35 (1959).
[47] Ripley B.D. *Spatial Statistics*, John Wiley and Sons, New York (1981).
[48] Diggle P.J. *Statistical Analysis of Spatial Point Patterns*, Oxford University Press, Oxford (2003).
[49] Oliver P., Cavarero M. Comparison between longitudinal tensile characteristics of thin and thick thermoset composite laminates: influence of curing conditions, *Computers and Structures*, 76, 125–37 (2000).
[50] Liu L., Zhang B.-M., Wang D.-F., Wu Z.-J. Effects of cure cycles on void content and mechanical properties of composite laminates, *Composite Structures*, 73, 303–9 (2006).
[51] Park H.C., Goo N.S., Min K.J., Yoon K.J. Three-dimensional cure simulation of composite structures by the finite element method, *Composite Structures*, 62, 51–7 (2003).

[52] Yi S., Hilton H.H., Ahmad M.F.A. A finite element approach for cure simulation of thermosetting matrix composites, *Computers and Structures*, 64, 383–8 (1996).

[53] Carlone P., Palazzo G.S. Thermo-chemical and rheological finite element analysis of the cure process of thick carbon–epoxy composite laminates, *International Journal of Material Forming*, 2, 137–40 (2009).

[54] Costa V.A.F., Sousa A.C.M. Modeling of flow and thermo-kinetics during the cure of thick laminated composites, *International Journal of Thermal Science*, 42, 15–22 (2003).

[55] Young W.-B. Compacting pressure and cure cycle for processing of thick composite laminates, *Composites Science Technology*, 54, 299–306 (1995).

[56] Carlone P., Palazzo G.S. Composite laminates cure cycle optimisation by meta-heuristic algorithms, *International Journal of Materials Product Technology*, 46, 106–23 (2013).

[57] Lee W.J., Loos A.C., Springer G.S. Heat of reaction degree of cure and viscosity of Hercules 3501-6 resin, *Journal of Composite Materials*, 16, 510–20 (1982).

[58] Kim J., Moon T.J., Howell J.R. Cure kinetic model, heat of reaction, and glass transition temperature of AS4/3501-6 graphite–epoxy prepregs, *Journal of Composite Materials*, 36, 2479–98 (2002).

[59] Sihn S., Roy A.K. Micromechanical analysis for transverse thermal conductivity of composites, *Journal of Composite Materials*, 45(11), 1245–55 (2011).

[60] Carlone P., Palazzo G.S. A micro-scale model for fiber tow characterization under nondeterministic assumption: longitudinal and transverse permeability, *Key Engineering Materials*, 554–557, 2348–54 (2013).

[61] Carlone P., Palazzo G.S. A multi-scale non-deterministic approach to composite curing process simulation, *Advanced Materials Research*, 750–752, 11–15 (2013).

[62] Liu X.L., Hillier W. Heat transfer and cure analysis for the pultrusion of a fiberglass–vinyl ester I beam, *Composite Structures*, 47, 581–8 (1999).

[63] Valliappan M., Roux J.A., Vaughan J.G. Die and post-die temperature and cure in graphite/epoxy composites, *Composites Part B: Engineering*, 27B, 1–9 (1996).

[64] Starr T.F. *Pultrusion for Engineers*, Woodhead (2000).

[65] Liu X.L., Crouch I.G., Lam Y.C. Simulation of heat transfer and cure in pultrusion with a general-purpose finite element package, *Composites Science and Technology*, 60, 857–64 (2000).

[66] Carlone P., Palazzo G.S., Pasquino R. Pultrusion manufacturing process development by computational modelling and methods, *Mathematical and Computer Modelling*, 44, 701–9 (2006).

[67] Baran I., Tutum C.C., Hattel J.H. The effect of thermal contact resistance on the thermosetting pultrusion process, *Composites Part B: Engineering*, 95, 995–1000 (2013).

[68] Joshi S.C., Lam Y.C. Three-dimensional finite-element/nodal-control-volume simulation of the pultrusion process with temperature-dependent

material properties including resin shrinkage, *Composites Science and Technology*, 61, 1539–47 (2001).

[69] Chachad Y.R., Roux J.A., Vaughan J.G. Three-dimensional characterization of pultruded fibreglass–epoxy composite materials, *Journal of Reinforced Plastics and Composites*, 14, 495–512 (1995).

[70] Chachad Y.R., Roux J.A., Vaughan J.G., Arafat E.S. Manufacturing model for three-dimensional irregular pultruded graphite/epoxy composites, *Composites*, 27, 201–10 (1996).

[71] Suratno B.R., Ye L., Mai Y.W. Simulation of temperature and curing profiles in pultruded composite rods, *Composites Science and Technology*, 58, 191–7 (1998).

[72] Li J., Joshi S.C., Lam Y.C. Curing optimization for pultruded composite sections, *Composites Science and Technology*, 62, 457–67 (2002).

[73] Joshi S.C., Lam Y.C., Tun U.W. Improved cure optimization in pultrusion with pre-heating and die-cooler temperature, *Composites*, 34, 1151–59 (2003).

[74] Baran I., Tutum C.C., Hattel J.H. Thermo-chemical modelling strategies for the pultrusion process, *Applied Composite Materials*, 20(6), 1247–63 (2013), doi:10.1007/s10443-013-9331-x.

[75] Raper K.S., Roux J.A., McCarty T.A., Vaughan J.G. Investigation of the pressure behaviour in a pultrusion die for glass-fibre/epoxy composites, *Composites: Part A – Applied Science and Manufacturing*, 30, 1123–32 (1999).

[76] Gadam S.U.K., Roux J.A., McCarty T.A., Vaughan J.G. The impact of pultrusion processing parameters on resin pressure rise inside a tapered cylindrical die for glass-fibre/epoxy composites, *Composites Science and Technology*, 60, 945–58 (2000).

[77] Price H.L., Cupschalk S.G. Pulling force and its variation in composite materials pultrusion, in Han C.D. (ed.), *Polymer Blends and Composites in Multiphase Systems*, ACS Publications, pp. 301–22 (1984).

[78] Lackey E., Vaughan J.G. An analysis of factors affecting pull force for the pultrusion of graphite/epoxy composites, *Journal of Reinforced Plastics and Composites*, 13, 188–99 (1994).

[79] Carlone P., Palazzo G.S. Viscous pull force evaluation in the pultrusion process by a finite element thermo-chemical rheological model, *International Journal of Material Forming*, 1, 831–4 (2008).

[80] Srinivasagupta D., Potaraju S., Kardos J.L., Joseph B. Steady state and dynamic analysis of a bench-scale injected pultrusion process, *Composites: Part A – Applied Science and Manufacturing*, 34, 835–46 (2003).

[81] Carlone P., Palazzo G.S. Computational modeling of the pulling force in a conventional pultrusion process, *Advanced Materials Research*, 772, 399–406 (2013).

[82] Moschiar S.M., Reboredo M.M., Larrondo H., Vazquez A. Pultrusion of epoxy matrix composites: pulling force model and thermal stress analysis, *Polymer Composites*, 17(6), 850–8 (1996).

[83] Yun M.S., Lee W.I. Analysis of pulling force during pultrusion process of phenolic foam composites, *Composites Science and Technology*, 68, 140–6 (2008).

[84] Carlone P., Baran I., Hattel J.H., Palazzo G.S. Computational approaches for modeling the multiphysics in pultrusion process, *Advances in Mechanical Engineering*, 2013, Article ID 301875, 1–14 (2013).

[85] Douglas Jr J. On the numerical integration of $u_{xx}+u_{yy}=u_t$ by implicit methods, *Journal of the Society for Industrial and Applied Mathematics*, 3(1), 45–65 (1955).

[86] Peaceman D.W., Rachford Jr H.H. The numerical solution of parabolic and elliptic differential equations, *Journal of the Society for Industrial and Applied Mathematics*, 3(1), 28–41 (1955).

[87] *ANSYS CFX 13.0 Reference Guide*, ANSYS, Inc. (2010).

[88] Baran I., Tutum C.C., Nielsen M.W., Hattel J.H. Process induced residual stresses and distortions in pultrusion, *Composites Part B: Engineering*, 51, 148–61 (2013).

[89] Trochu F., Ruiz E., Achim V., Soukane S. Advanced numerical simulation of liquid composite molding for process analysis and optimization, *Composites Part A: Applied Science and Manufacturing*, 37, 890–902 (2006).

[90] Song Y.S. Mathematical analysis of resin flow through fibrous porous media, *Applied Composite Materials*, 13(5), 335–43 (2006).

[91] Andersson H.M., Lundström T.S., Langhans N. Computational fluid dynamics applied to the vacuum infusion process, *Polymer Composites*, 26(2), 231–9 (2005).

[92] Joubaud L., Achim V., Trochu F. Numerical simulation of resin infusion and reinforcement consolidation under flexible cover, *Polymer Composites*, 26(4), 427 (2005).

[93] Ouahbi T., Saouab A., Bréard J., Ouagne P., Chatel S. Modelling of hydro-mechanical coupling in infusion process, *Composites Part A: Applied Science and Manufacturing*, 38, 1646–54 (2007).

[94] Kessels J.F.A., Jonker A.S., Akkerman R. Fully 2½D flow modelling of resin infusion under flexible tooling using unstructured meshes and wet and dry compaction properties, *Composites Part A: Applied Science and Manufacturing*, 38, 51–60 (2007).

[95] Wang Y., Grove S.M. Modelling microscopic flow in woven fabric reinforcements and its application in dual-scale resin infusion modelling, *Composites Part A: Applied Science and Manufacturing*, 39, 843–55 (2008).

[96] Bayldon J.M., Daniel I.M. Flow modeling of the VARTM process including progressive saturation effects, *Composites Part A: Applied Science and Manufacturing*, 40, 1044–52 (2009).

[97] Carlone P., Palazzo G.S. Computational analysis of the vacuum infusion process of reactive thermosetting resin, *Proceedings of the 7th International Conference on Engineering Computational Technology, ECT 2010*, Paper 64, doi:10.4203/ccp.94.64 (2010).

[98] Govignon Q., Bickerton S., Kelly P.A. Simulation of the reinforcement compaction and resin flow during the complete resin infusion process, *Composites Part A: Applied Science and Manufacturing*, 41, 45–57 (2010).

[99] Modi D., Correia N., Johnson M., Long A., Rudd C., Robitaille F. Active control of the vacuum infusion process, *Composites Part A: Applied Science and Manufacturing*, 38, 1271–87 (2007).

[100] Danisman M., Tuncol G., Kaynar A., Sozer E.M. Monitoring of resin flow in the resin transfer molding (RTM) process using point-voltage sensors, *Composites Science and Technology*, 67, 367–79 (2007).

[101] Luthy T., Ermanni P. Linear direct current sensing system for flow monitoring in liquid composite moulding, *Composites Part A: Applied Science and Manufacturing*, 33, 385–97 (2002).

[102] Vaidya U.K., Jadhav N.C., Hosur M.V., Gillespie Jr J.W., Fink B.K. Assessment of flow and cure monitoring using direct current and alternating current sensing in vacuum-assisted resin transfer molding, *Smart Materials Structures*, 9, 727–36 (2000).

[103] Tuncol G., Danisman M., Kaynar A., Sozer E.M. Constraints on monitoring resin flow in the resin transfer molding (RTM) process by using thermocouple sensors, *Composites Part A: Applied Science and Manufacturing*, 38, 1363–86 (2007).

[104] Lekakou C., Cook S., Deng Y., Ang T.W., Reed G.T. Optical fibre sensor for monitoring flow and resin curing in composite manufacturing, *Composites Part A: Applied Science and Manufacturing*, 37, 934–8 (2006).

[105] Schmachtenberg E., Schulte zur Heide J., Topker J. Application of ultrasonics for the process control of resin transfer moulding (RTM), *Polymer Testing*, 24, 330–8 (2005).

[106] Yenilmez B., Sozer E.M. A grid of dielectric sensors to monitor mold filling and resin cure in resin transfer molding, *Composites Part A: Applied Science and Manufacturing*, 40, 476–89 (2009).

[107] Carlone P., Palazzo G.S. Flow monitoring and permeability measurements in LCM processes by the means of a dielectric sensor, *Key Engineering Materials*, 504–506, 289–94 (2012).

[108] Gebart B.R. Permeability of unidirectional reinforcements for RTM, *Journal of Composite Materials*, 26(8), 1100–33 (1992).

[109] Choi M.A., Lee M.H., Chang J., Lee S.J. Permeability modeling of fibrous media in composite processing, *Journal of Non-Newtonian Fluids*, 79, 585–98 (1998).

[110] Carlone P., Palazzo G.S. A micro-scale model for fiber tow characterization under nondeterministic assumption: longitudinal and transverse permeability, *Key Engineering Materials*, 554–557, 2348–54 (2013).

[111] Grujicic M., Chittajallu K.M., Walsh S. Non-isothermal perform infiltration during the vacuum-assisted resin transfer molding (VARTM) process, *Applied Surface Science*, 245, 51–64 (2005).

[112] Grimsley B.W. Characterization of the vacuum assisted resin transfer molding process for fabrication of aerospace composites, Master's thesis, Virginia Polytechnic Institute and State University (2005).

[113] Johnson R.J., Pitchumani R. Enhancement of flow in VARTM using localized induction heating, *Composites Science and Technology*, 63, 2201–15 (2003).

[114] Wang Y., Grove S.M. Modelling microscopic flow in woven fabric reinforcements and its application in dual-scale resin infusion modelling, *Composites Part A: Applied Science and Manufacturing*, 39, 843–55 (2008).

[115] Zhou F., Alms J., Advani S.G. A closed form solution for flow in dual scale fibrous porous media under constant injection pressure conditions, *Composites Science and Technology*, 68, 699–708 (2008).

[116] Kuentzer N., Simacek P., Advani S.G., Walsh S. Permeability characterization of dual scale fibrous porous media, *Composites Part A: Applied Science and Manufacturing*, 37, 2057–68 (2006).

[117] Kolda T.G., Lewis R.M., Torczon V. Optimization by direct search: new perspectives on some classical and modern methods, *SIAM Review*, 45(3), 385–482 (2003).

[118] Hvattum L.M., Glover F. Finding local optima of high-dimensional functions using direct search methods, *European Journal of Operational Research*, 195(1), 31–45 (2009).

[119] Wilcox J.A.D., Wright D.T. Towards pultrusion process optimisation using artificial neural networks, *Journal of Materials Processing Technology*, 83, 131–41 (1998).

[120] Carlsson A., Åström T. Experimental investigation of pultrusion of glass fibre reinforced polypropylene composites, *Composites Part A: Applied Science and Manufacturing*, 29A, 585–93 (1998).

[121] Haj-Ali R., Kilic H. Nonlinear behavior of pultruded FRP composites, *Composites Part B: Engineering*, 33, 173–91 (2002).

[122] Coelho R.M.L., Calado V.M.A. An optimization procedure for the pultrusion process based on a finite element formulation, *Polymer Composites*, 23, 329–41 (2002).

[123] Baran I., Tutum C.C., Hattel J.H. Optimization of the thermosetting pultrusion process by using hybrid and mixed integer genetic algorithms, *Applied Composite Materials*, 20(4), 449–63 (2013).

[124] Chen X.K., Jiang B.Y., Zhou Z. Optimization of CFRP pultrusion process with NSGA-II and ANN, *Advanced Materials Research*, 538–541, 2705–11 (2012).

[125] Santos L.S., Biscaia Jr E.C., Pagano R.L., Calado V.M.A. CFD-optimization algorithm to optimize the energy transport in pultruded polymer composites, *Brazilian Journal of Chemical Engineering*, 29(3), 559–66 (2012).

[126] Carlone P., Palazzo G.S., Pasquino R. Pultrusion manufacturing process development: Cure optimization by hybrid computational methods, *Computers and Mathematics with Applications*, 53(9), 1464–71 (2007).

[127] Carlone P., Palazzo G.S. Cure optimization in pultrusion process by a heuristic-analytical approach, *ASME International Mechanical Engineering Congress and Exposition, IMECE2006*, Chicago, 5–10 November 2006, Code 69689 (2006).

[128] Goldberg D.E. *Genetic Algorithms in Search, Optimization and Machine Learning*, Addison-Wesley, New York (1989).

[129] Coley D. *An Introduction to Genetic Algorithms for Scientists and Engineers*, World Scientific, Singapore (2003).

[130] Spendley W., Hext G.R., Himsworth F.R. Sequential application of simplex designs in optimisation and evolutionary operation, *Technometrics*, 4(4), 441–61 (1962).

[131] Nelder J.A., Mead R. A simplex method for function minimization, *The Computer Journal*, 7(4), 308–13 (1965).

[132] Cardoso M.F., Salcedo R.L., Feyo de Azevedo S. The simplex-simulated annealing approach to continuous non-linear optimization, *Computers & Chemical Engineering*, 20(9), 1065–80 (1996).

[133] Lagarias J.C., Reeds J.A., Wright M.H., Wright P.E. Convergence properties of the Nelder–Mead simplex method in low dimensions, *SIAM Journal on Optimization*, 9(1), 112–47 (1998).

[134] Luersen M.A., Le Riche R. Globalized Nelder–Mead method for engineering optimization, *Computers & Structures*, 82(23–26), 2251–60 (2004).

[135] Mawardi A., Pitchumani R. Optimal cure and current cycles for curing of composites using embedded resistive heating elements, *Journal of Manufacturing Science and Engineering*, 122(1), 124–31 (2000).

[136] Bailleul J.-L., Sobotka V., Delaunay D., Jarny Y. Inverse algorithm for optimal processing of composite materials, *Composites Part A: Applied Science and Manufacturing*, 34(8), 695–708 (2003).

[137] Dufour P., Michaud D.J., Tourè Y., Dhurjati P.S. A partial differential equation model predictive control strategy: application to autoclave composite processing, *Computers & Chemical Engineering*, 28(4), 545–56 (2004).

[138] Carlone P., Palazzo G.S. Computational optimization of the thermal cure cycle of composite laminates, *Proceedings of the 7th International Conference on Engineering Computational Technology*, Valencia, Spain (2010).

[139] Carlone P., Palazzo G.S. A simulation based metaheuristic optimization of the thermal cure cycle of carbon–epoxy composite laminates, *Proceedings of the 14th International Esaform Conference on Material Forming*, AIP Conference Proceedings, Belfast, United Kingdom, pp. 5–10 (2011).

[140] Ruiz E., Trochu F. Multi-criteria thermal optimization in liquid composite molding to reduce processing stresses and cycle time, *Composites Part A: Applied Science and Manufacturing*, 37(6), 913–24 (2006).

[141] Kam T.Y., Lai F.M., Sher H.F. Optimal parameters for curing graphite/epoxy composite laminates, *Journal of Materials Processing Technology*, 48(1–4), 357–63 (1995).

[142] *ANSYS Parametric Design Language Guide*, Ansys, Inc.

[143] Kirkpatrick S., Gelatt Jr C.D., Vecchi M.P. Optimization by simulated annealing, *Science*, 220(4598), 671–80 (1983).

[144] Blau P.J. Composition, functions and testing of friction brake materials and their additives, Oak Ridge National Laboratory, ORNL/TM-2001/64 (2001).

[145] Jang H., Kim S.J. The effects of antimony trisulfide (Sb_2S_3) and zirconium silicate ($ZrSiO_4$) in automotive brake friction material on friction characteristics, *Wear*, 239, 229–36 (2000).

[146] Kim S.J., Jang H. Friction characteristics of non-asbestos organic and low-steel friction materials: The comparative study, *KSTLE International Journal*, 1(1), 1–7 (2000).

[147] Eriksson M. Friction and contact phenomena of disc brakes related to squeal, Doctoral dissertation, Faculty of Science and Technology, University of Uppsala (2000).

[148] Cho M.H., Kim S.J., Basch R.H., Fash J.W., Jang H. Tribological study of cast iron with automotive brake linings: The effect of rotor microstructure, *Tribology International*, 36, 537–45 (2003).

[149] Eriksson M., Jacobson S. Tribological surfaces of organic brake pads, *Tribology International*, 33, 817–27 (2000).

[150] Sasaki Y. Development philosophy of friction materials for automotive disc brakes, *Proceedings of the Japan Society of Automotive Engineers*, 1995, No. 9531679, 407–12 (1995).

[151] Thuresson D. Influence of material properties on sliding contact-braking applications, *Wear*, 257(5–6), 451–60 (2004).

[152] Jacko M.G., Tsang P.H.S., Rhee S.K. Wear debris compactions and friction film formation of polymer composites, *Wear*, 133(1), 23–38 (1989).

[153] Kim S.J., Cho M.H., Lim D.-S., Jang H. Synergistic effects of aramid pulp and potassium titanate whiskers in automotive friction material, *Wear*, 251, 1484–91 (2001).

[154] Bahadur S. The development of transfer layers and their role in polymer tribology, *Wear*, 245, 92–9 (2000).

[155] Cho M., Bae E., Jang H., Jeong G., Kim D., Choi C. The role of raw material ingredients of brake linings on the formation of transfer film and friction characteristics, SAE Technical Paper 2001-01-3130 (2001), doi:10.4271/2001-01-3130.

[156] Eggleston D., Whitaker R. A fundamental tribochemical study of third body layered formed during automotive friction braking, *Wear*, 179(1–2), 75–81 (1994).

[157] Kim S.J., Jang H. Friction and wear of friction materials containing two different phenolic resins reinforces with aramid pulp, *Tribology International*, 33, 477–84 (2000).

[158] Jang H., Ko K., Kim S.J., Basch R.H., Fash J.W. The effect of metal fibers on the friction performance of automotive brake friction materials, *Wear*, 256(3–4), 406–14 (2004).

[159] Eriksson M., Lord J., Jacobson S. Wear and contact conditions of brake pads: Dynamical in situ studies of pad on glass, *Wear*, 249, 272–8 (2001).

[160] Jang H., Lee J.S., Fash J.W. Compositional effects of the brake friction material on creep groan phenomena, *Wear*, 251, 1477–83 (2001).

[161] Satapathy B.K., Bijwe J. Performance of friction materials based on variation in nature of organic fibers – Part II. Optimisation by balancing and ranking using multiple criteria decision model (MCDM), *Wear*, 257, 585–9 (2004).

[162] Ostermeyer G.P. On the dynamics of the friction coefficient, *Wear*, 254, 852–8 (2003).

[163] Jintang G. Trobochemical effects in formation of polymer transfer film, *Wear*, 245, 100–6 (2000).

[164] Zhang Z., Reincke R., Klein P., Friedrich K., Velten K. Wear of polymer composites using artificial neural networks, *International Conference on Composites in Material and Sstructural Engineering, CMSE*, Prague (2001).

[165] Zhang Z., Klein P., Friedrich K. Dynamic mechanical properties of PTFE based short carbon fibre reinforced composites: experiment and artificial neural network prediction, *Composite Science and Technology*, 62, 1001–9 (2002).
[166] Zhang Z., Klein P., Friedrich K. Artificial neural networks applied to polymer composites: a review, *Composite Science and Technology*, 63, 2029–44 (2003).
[167] Zhang Z., Barkolula N.-M., Karger-Kocsic J., Friedrich K. Artificial neural network predictions on erosive wear polymers, *Wear*, 255, 708–13 (2003).
[168] Krose B., van der Samgt P. *An Introduction to Neural Networks*, University of Amsterdam (1996).
[169] Devabhaktuni V.K., Yagoub M.C.E., Fang Y., Xu J., Zhang Q.J. *Neural Networks for Microwave Modelling: Model Development Issues and Nonlinear Modelling Techniques*, John Wiley and Sons (2001).
[170] Aleksendrić D., Duboka Č. Prediction of automotive friction material characteristics using artificial neural networks – cold performance, *Wear*, 261, 269–82 (2006).
[171] Kim S.J., Kim K.S., Jang H. Optimization of manufacturing parameters for a brake lining using Taguchi method, *Journal of Material Processing*, 136, 202–8 (2003).
[172] Zhang Z., Friedrich K., Velten K. Prediction on tribological properties of short fibre composites using artificial neural networks, *Wear*, 252, 668–75 (2002).
[173] *DARPA Neural Network Study*, AFCEA International Press, p. 60 (1998).
[174] Demuth H., Beale M. *Neural Network Toolbox for Use with MATLAB, Users Guide*, ver. 4.0., The MathWorks, Inc. (1998).
[175] Kolich M., Seal N., Taboun S. Automobile seat comfort prediction: statistical model vs. artificial neural network, *Applied Ergonomics*, 35, 275–84 (2004).
[176] Liu D., Chang T.S., Zhang Y. A constructive algorithm for feedforward neural networks with incremental training, *IEEE Transactions on Circuits and Systems*, 49(12), 1876–9 (2002).
[177] Fonseca D., Navaresse D.O, Moynihan G.P. Simulation metamodeling through artificial neural networks, *Engineering Application of Artificial Intelligence*, 16, 177–83 (2003).
[178] Takayama K., Fujikawa M., Obata Y., Morishita M. Neural network based optimization of drug formulations, *Advanced Drug Delivery Reviews*, 55, 1217–31 (2003).
[179] Peng K., Ge S.S., Wen C. An algorithm to determine neural network hidden layer size and weight coefficients, *Proceedings of the 15th IEEE International Symposium on Intelligent Control*, Rio Patras, Greece (2000).
[180] Windrow B., Lehr M. 30 years of adaptive neural networks: perceptron, Madaline, and backpropagation, *Proceedings of the IEEE*, 78(9), 1415–42 (1990), doi:10.1109/5.58323.
[181] Aleksendrić D. Artificial neural networks in automotive brakes' friction material development, Doctoral thesis, Faculty of Mechanical Engineering, University of Belgrade (2007).

[182] Cogan B. The power of neural networks, Review, *Scientific Computing World*, March/April (2003).

[183] Redondo M.F., Espinoza C.H. Weight initialization methods for multilayer feedforward, *Proceedings of ESANN 2001, European Symposium on Artificial Neural Networks*, Bruges, Belgium, pp. 119–124 (2001).

[184] MacKay D.J.C. *Bayesian Methods for Neural Networks: Theory and Applications*, University of Cambridge (1995).

[185] Jang H., Lee J., Kim S., Jung K. The effect of solid lubricants on friction characteristics, SAE Technical Paper 982235 (1998), doi:10.4271/982235.

[186] Decuzzi P., Demelio G. The effect of material properties on the thermoelastic stability of sliding systems, *Wear*, 252, 311–21 (2002).

[187] Eriksson M., Bergman F., Jacobson S. On the nature of tribological contact in automotive brakes, *Wear*, 252, 26–36 (2002).

[188] Satapathy B.K, Bijwe J. Performance friction materials based on variation in nature of organic fibers – Part I. Fade and recovery behaviour, *Wear*, 257, 573–84 (2004).

[189] Myshkin N.K. Friction transfer film formation in boundary lubrication, *Wear*, 245, 116–24 (2000).

[190] Nichols K. Getting engineering changes under control, *Journal of Engineering Design*, 1(1), 5–15 (1990).

[191] Aleksendrić D., Duboka C. Fade performance prediction of automotive friction materials by means of artificial neural networks, *Wear*, 262(7–8), 778–90 (2007).

[192] Voracek J. *Introduction to Knowledge-Based Intelligent Systems*, Pearson Education (2002).

[193] Demuth H., Beale M. *Neural Network Toolbox User's Guide for Use with MATLAB*, The MathWorks, Inc. (2002).

[194] Fillot F., Iordanoff I., Berthier Y. Wear modelling and the third body concept, *Wear*, 949–57 (2007).

[195] Kato K. Wear in relation to friction – a review, *Wear*, 241, 151–7 (2000).

[196] Bahadur S. The development of transfer layers and their role in polymer tribology, *Wear*, 245, 92–9 (2000).

[197] Archard J.F. Contact and rubbing of flat surfaces, *Journal of Applied Physics*, 24, 981–8 (1953).

[198] Rhee S.K. Wear equation for polymers sliding against metal surfaces, *Wear*, 16, 431–45 (1970).

[199] Ramalhoa A., Miranda J.C. The relationship between wear and dissipated energy in sliding systems, *Wear*, 260, 361–7 (2006).

[200] AbuBakar A.R., Ouyang H. Wear prediction of friction material and brake squeal using the finite element method, *Wear*, 264, 1069–76 (2008).

[201] Talib R.J., Muchtar A., Azhari C.H. Microstructural characteristics on the surface and subsurface of semimetallic automotive friction materials during braking process, *Journal of Materials Processing Technology*, 140, 694–9 (2003).

[202] Velten K., Reincke R., Friedrich K. Wear volume prediction with artificial neural network, *Tribology International*, 33, 731–6 (2000).

[203] Zhang Z., Friedrich K. Artificial neural networks applied to polymer composites: a review, *Composites Science and Technology*, 63, 2029–44 (2003).

[204] Aleksendrić D. *Modelling of Brake Friction Material Performance at Elevated Temperatures*, Chandos, Oxford, pp. 13–20 (2009).

[205] Aleksendrić D., Duboka Č. Artificial technologies in sustainable braking system development, *International Journal of Vehicle Design*, 46, 237–49 (2008).

[206] Aleksendrić D., Barton D. Neural network prediction of disk brake performance, *Tribology International*, 42(7), 1074–80 (2007).

[207] Aleksendrić D., Duboka Č., Mariotti G.V. Neural modelling of friction material cold performance, *Proceedings of the Institution of Mechanical Engineers Part D: Journal of Automobile Engineering*, 222(7), 1021–9 (2008).

[208] Aleksendrić D. Neural network prediction of brake friction materials wear, *Wear*, 268(1–2), 117–25 (2010).

[209] Jiang Z., Zhang Z., Friedrich K. Prediction on wear properties of polymer composites with artificial neural network, *Composite Science and Technology*, 67, 168–76 (2007).

[210] Todorovic J., Duboka C., Arsenic Z. Operational life expectancy of rubbing elements in automotive brakes, *Tribology International*, 28, 423–32 (1995).

[211] Gurunath P.V., Bijwe J. Friction and wear studies on brake-pad materials based on newly developed resin, *Wear*, 263, 1212–19 (2007).

[212] Satapathy B.K., Bijwe, J. Performance friction materials based on variation in nature of organic fibers – Part I. Fade and recovery behavior, *Wear*, 257, 573–84 (2004).

[213] Seong J.K., Kwang S.K., Ho J. Optimization of manufacturing parameters for a brake lining using Taguchi method, *Journal of Materials Processing Technology*, 136, 202–8 (2003).

[214] Schiffner K., Heftrich M., Brecht J. Modeling of compaction processes of friction material mixes, SAE Technical Paper 2002-01-2594 (2002), doi:10.4271/2002-01-2594.

[215] Han L., Huang L., Zhang J., Lu Y. Optimization of ceramic friction materials, *Composites Science and Technology*, 66, 2895–906 (2006).

[216] Lu Y., Tang C.F., Wright M. Optimization of a commercial brake pad formulation, *Journal of Applied Polymer Science*, 84, 2498–504 (2002).

[217] Yun R., Martynková S., Lu Y. Performance and evaluation of nonasbestos organic brake friction composites with SiC particles as an abrasive, *Journal of Composite Materials*, 45, 1585–93 (2011).

[218] Matejka V., Martynková G., Ma Y., Lu Y. Semimetallic brake friction materials containing $ZrSiO_4$: friction performance and friction layers evaluation, *Journal of Composite Materials*, 43, 1421–34 (2009).

[219] Cetinel H., Ozturk H., Celik E., Karlik B. Artificial neural network-based prediction technique for wear loss quantities in Mo coatings, *Wear*, 261, 1064–8 (2006).

[220] Gyurova L., Friedrich K. Artificial neural networks for predicting sliding friction and wear properties of polyphenylene sulfide composites, *Tribology International*, 44, 603–9 (2011).

[221] Senatore A., D'Agostino V., Di Giuda R., Petrone V. Experimental investigation and neural network prediction of brakes and clutch material frictional behaviour considering the sliding acceleration influence, *Tribology International*, 44, 1199–1207 (2011).

[222] Aleksendrić D., Senatore A. Optimization of manufacturing process effects on brake friction material wear, *Journal of Composite Materials*, 46(22), 2777–91 (2012). (Online version available at: *http://jcm.sagepub.com/content/46/22/2777.*)

[223] Day A.J., Tirovic M., Newcomb T.P. Thermal effects and pressure distributions in brakes, *Proceedings of the Institution of Mechanical Engineers, Part D: Journal of Automobile Engineering*, 205, 199–205 (1991).

[224] Blanco A., Delgado M., Pegalajar M.C. A real-coded genetic algorithm for training recurrent neural networks, *Neural Networks*, 14, 93–105 (2001).

[225] Thompson G.J., Atkinson C.M., Clark N.N., Long T.W., Hanzevack E. Neural network modelling of the emissions and performance of a heavy-duty diesel engine, *Proceedings of the Institution of Mechanical Engineers, Part D: Journal of Automobile Engineering*, 214, 111–26 (2000).

[226] Manan A., Vio G.A., Harmin M.Y., Cooper J.E. Optimization of aeroelastic composite structures using evolutionary algorithms, *Engineering Optimization*, 42, 171–84 (2009).

[227] Yazdanmehr M., Mousavi Anijdan S.H., Bahrami A. Using GA–ANN algorithm to optimize soft magnetic properties of nanocrystalline mechanically alloyed Fe–Si powders. *Computational Materials Science*, 44, 1218–21 (2009).

[228] Mashinchi M.R., Selamat A. An improvement on genetic-based learning method for fuzzy artificial neural networks, *Applied Soft Computing*, 9, 1208–16 (2009).

[229] Ko Y.D., Moon P., Kim C.E., Ham M.H., Myoung J.M., Yun I. Modelling and optimization of the growth rate for ZnO thin films using neural networks and genetic algorithms. *Expert Systems with Applications*, 36, 4061–6 (2009).

[230] Sedki A., Ouazar D., El Mazoudi E. Evolving neural network using real coded genetic algorithm for daily rainfall-runoff forecasting, *Expert Systems with Appl*ications, 36, 4523–7 (2009).

[231] Sellgren U., Söderberg A. Robust brake-feel design, *Proceedings of the NAFEMS World Congress*, Hersonissos, Crete, Greece (2009).

[232] Day A., Ho H., Hussain K., Johnstone A. Brake system simulation to predict brake pedal feel in a passenger car, SAE Technical Paper 2009-01-3043 (2009), doi:10.4271/2009-01-3043.

[233] Basch R., Sanders P., Hartsock D., Evans C. Correlation of lining properties with brake pedal feel, SAE Technical Paper 2002-01-2602 (2002), doi:10.4271/2002-01-2602.

[234] Aleksendrić D., Cirovic V., Sovrović V. Dynamic control of disc brake performance, *SAE International Journal of Passenger Cars – Mechanical Systems*, 5(4), 1266–72 (2012), doi:10.4271/2012-01-1837.
[235] Ćirović V., Aleksendrić D. Dynamic modelling of disc brake contact phenomena, *FME Transactions*, 39, 177–83 (2011).
[236] Aleksendric D., Duboka C., Cirovic V. Intelligent control of disc brake operation, SAE Technical Paper 2008-01-2570 (2008), doi:10.4271/2008-01-2570.
[237] Ćirović V., Aleksendrić D. Intelligent control of automotive braking system, *FISITA 2008 World Automotive Congress*, Munich, F2008-SC-046 (2008).
[238] Ćirović V., Aleksendrić D. Development of neural network model of disc brake operation, *FME Transactions*, 38, 29–38 (2010).
[239] Xiao G., Zhu Z. Friction materials development by using DOE/RSM and artificial neural network, *Tribology International*, 43, 218–27 (2010).
[240] Rukiye E., Nurettin Y. An experimental study of the effects of manufacturing parameters on the tribological properties of brake lining, *Wear*, 268, 1524–32 (2010).
[241] Aleksendric D., Cirovic V. Effect of friction material manufacturing conditions on its wear, SAE Technical Paper 2010-01-1679 (2010), doi:10.4271/2010-01-1679.
[242] Aleksendrić D. An inverse neural network model of disc brake performance at elevated temperatures, in Flores A.J. (ed.), *Focus on Artificial Neural Networks*, pp. 151–70, Nova Science Publishers, New York (2011).
[243] Aleksendrić D. Manufacturing of brake friction materials, in Savarese A.B. (ed.), *Manufacturing Engineering*, pp. 89–108, Nova Science Publishers, New York (2013).
[244] Aleksendrić D., Barton D.C., Vasić B. Prediction of brake friction materials recovery performance using artificial neural networks, *Tribology International*, 43, 2092–9 (2010).
[245] Aleksendrić D., Jakovljević Ž., Ćirović V. Intelligent control of braking process, *Expert Systems with Applications*, 39, 11758–65 (2012).
[246] Ćirović V., Aleksendrić D., Mladenović D. Braking torque control using recurrent neural networks, *Proceedings of the Institution of Mechanical Engineers, Part D: Journal of Automobile Engineering*, 226, 754–66 (2012).
[247] Ćirović V., Aleksendrić D. Adaptive neuro-fuzzy wheel slip control, *Expert Systems with Applications*, 40(13), 5197–209 (2013).
[248] Ćirović V., Aleksendrić D., Smiljanić D. Longitudinal wheel slip control using dynamic neural networks, *Mechatronics*,23, 135–46 (2013).
[249] Aleksendrić D. Intelligent braking – technology, performance and economic challenge, in Zajac M., Nowaczek R. (eds), *Airports and the Automotive Industry: Security Issues, Economic Efficiency and Environmental Impact*, pp. 33–64, Nova Science Publishers, New York (2013).
[250] Simon J.R., Cole D. Modelling nonlinear vehicle dynamics with neural networks, *International Journal of Vehicle Design*, 53, 260–87 (2010).

Index

Edwards Brothers Malloy
Thorofare, NJ USA
March 5, 2015